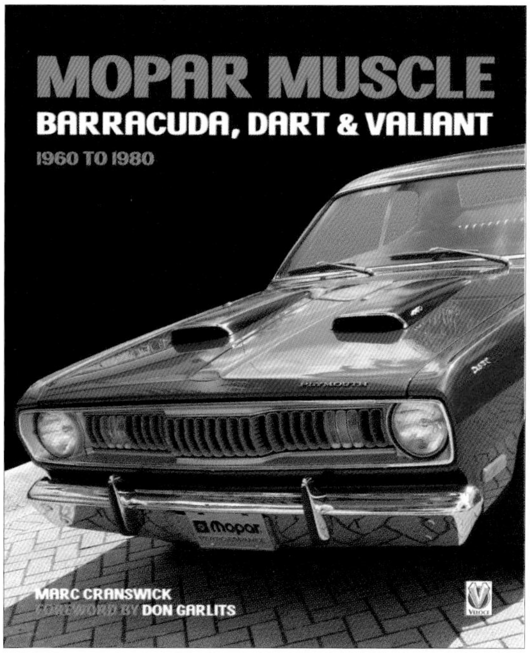
MOPAR MUSCLE
BARRACUDA, DART & VALIANT
1960 TO 1980
MARC CRANSWICK
FOREWORD BY DON GARLITS
Mopar

More great books from Veloce:

First published in March 2018 by Veloce Publishing Limited, Veloce House, Parkway Farm Business Park, Middle Farm Way, Poundbury, Dorchester DT1 3AR, England. Tel +44 (0)1305 260068 / Fax 01305 250479 / e-mail info@veloce.co.uk / web www.veloce.co.uk www.velocebooks.com. ISBN: 978-1-787110-71-7 UPC: 6-36847-01071-3.

Readers with ideas for automotive books, or books on other transport or related hobby subjects, are invited to write to the editorial director of Veloce Publishing at the above address. British Library Cataloguing in Publication Data – A catalogue record for this book is available from the British Library. Typesetting, design and page make-up all by Veloce Publishing Ltd on Apple Mac. Printed in India by Parksons Graphics.

MOPAR MUSCLE

BARRACUDA, DART & VALIANT

1960 TO 1980

MARC CRANSWICK

FOREWORD BY **DON GARLITS**

VELOCE

Contents

Foreword – Recollections by Don Garlits 5

Introduction – The economy car meets the muscle car 6

Chapter 1: Barracuda 1964-69 – The A-body Fish 7

Chapter 2: Barracuda & Challenger 1970-74 – The E-bodies 28

Chapter 3: Dart – From senior compact to muscle car 52

Chapter 4: Plymouth Valiant – The economy car King 87

Chapter 5: From F-body to Hellcat & Demon! – The return of high-performance 116

Appendix 149
- Mr Norm's Grand Spaulding Dodge 149
- Mopar Missiles & Hemi Colt 152
- Mopar motored V8 exotics 155
- Chrysler overseas 163
- Mopar's marvelous quartet 165
- Mopar items of interest 169
- Endnotes 170

Index 175

Foreword – Recollections by Don Garlits

The 1960s were probably the most exciting and innovative time in Drag Racing, especially for me! I set the NHRA Top Fuel Record at over 200mph, and Chrysler introduced the unbelievable 426 Hemi 'Brute,' which would become the engine of choice for all Top Fuel and Funny Cars, forever! On top of all that, Dodge and Plymouth introduced the 'Max-Wedge' Hi-Performance cars in 1962, which soon dominated the Drag Strips everywhere. Then came the 'Light Weights' in 1963, and the Hemi cars the next year. I got so busy that I moved my operation to Troy, Michigan – a small town north of Detroit – where I could be close for R&D for Dodge Racing. However, the Michigan winters soon sent me back to Florida, but the Chrysler/Mopar relationship survives to this day!

I don't think we will ever see a time like that again, where, every week, the factories and racers would show up at a strip somewhere, with something totally new. The current rules of the National associations just wouldn't allow it to happen again! This, in fact, goes for all Motorsports. It all got too fast and dangerous to allow those times to repeat themselves, so that wonderful era is gone forever. Chrysler/Mopar and myself played a very important role in that era. I am proud to have been a part of the '60s and especially proud to have such a long association with Chrysler/Mopar for nearly 60 years!

Don Garlits, CEO
Don Garlits Museum of Drag Racing Inc

Introduction – The economy car meets the muscle car

"If thou shouldst lay up even a little upon a little, and shouldst do this often, soon would even this become great"
– Hesiod, 720 BC

So it was with the 101 horsepower, six cylinder 1960 Plymouth Valiant. Chrysler Corp kept adding incrementally, until it created the 425 horse 1970 Plymouth Hemi 'Cuda. Technically still a compact, such a development – from plain economy car to earth-shattering, all-conquering, fire and brimstone V8 bolide – could only have transpired during America's muscle car era.

Chrysler Corp's compacts and senior compacts had the rare ability to have an equal image as both an economy car and a supercar. Others were seen as either one or the other, but Dart could be a Mobil Economy Run champ and Mr Norm special simultaneously – at least in the public's mind, if not in the one car. Similarly, Duster could be the ideal budget car, or pony car.

Image aside, Chrysler Corp engineering's torsion bar suspension, handling, and high output V8s gave A-body the substance to embarrass a Corvette, in a sober, street sleeper shape. When styling became key in the pony/muscle car wars, Chrysler Corp obliged with E-body 'Cuda and Challenger R/T. Supercars in looks, word, and deed; one didn't need to be plum purple crazy to think that!

In the serious business of racing, Chrysler Corp cars were the go-to machines for those wishing to win. Little wonder governing authorities tried to end the dominance of Mopar Muscle via the rulebook: few could do it on the strip or track. Even when racing was done, and Chrysler F-body Aspen and Volaré provided the luxury that America wanted, the engineers didn't give up. In any given year, Chrysler engineers tried their best to make F-body R/T and Road Runner put the heat on rivals.

Chrysler Corp survived challenging business times, and changes in corporate ownership, to rediscover rear drive V8 high-performance cars in the 21st century. New Letter Series cars, Chargers and Challengers now show that speed demon still lives at Highland Park.

Marc Cranswick

— Dedicated to Marie & Erika —

1

Barracuda 1964-69 – The A-body Fish

A sport compact with broad appeal

"It's not that kind of car." That was Chrysler's response when *Car and Driver* grilled its Mopar man on why Barracuda didn't have Mustang's myriad engine and handling options out of the starting gate. He might have been telling the truth: Barracuda had an almost utilitarian nature on release, and it was almost called the Panda!

Panda was indeed the first choice of Plymouth's executives. They really didn't seem to have a handle on the youth market. Fortunately, the design staff stood their ground, and chief stylist John Sampson's Barracuda suggestion won out. The coupe's initial title on debut was Plymouth Valiant Barracuda, and that wasn't false advertising.

The coupe's nameplate suggested its position as a sporty adjunct to the successful Valiant range. Living up to its name, and saving the engineering pennies, entailed a fastback body that owed much to Valiant – specifically, Valiant's hood, headlamp bezels, windshield, vent panes, quarter panels, doors, A-pillar and front bumper!

All this sat upon Valiant's 106in wheelbase, with chassis of independent, non-parallel control arms, torsion bars and tube shocks at the front. Rearwards lived a leafsprung live axle, damped by tube shocks. Plainly visible was the single biggest piece of glass fitted to an automobile at that point in time. It was jointly developed by Chrysler designers and Pittsburgh Plate Glass (PPG). The fish's fastback had a 2070in^2 window, affording the pony car driver that rare quality – excellent visibility. Plymouth claimed the tinted window filtered 33 per cent of the sun's spectrum, 45 per cent of UV rays and 35 per cent of infra-red. In August 1964, *Motor Trend*'s Bob McVay said it never got too toasty back there. However, more than a few said that the rear cargo area could have cooked a turkey, if Thanksgiving fell in summer!

Under that megapane lay a 7ft load area, security panel, and fold-down back seat. Said seat had straps on each side and chrome bar. Latching device and front seat guard rail were provided. The luggage bar stopped load items slamming into the front seats, and one could fold down the rear seat with one hand. Chrysler intimated to *Car and Driver* in May 1964 that it had an eye on wagon and Vette sales. A car of broad appeal, indeed.

A rare 1965 Barracuda with 225 Slant-6 and TorqueFlite autobox. When the Barracuda came out, just 10 per cent of buyers chose the 170/225 Slant-6s. (Courtesy Romaine Dennis, Good Time Classics goodtimeclassics.net)

At 2070in², the Barracuda's fastback glass pane was the biggest single piece of glass fitted to an automobile at that time. (Courtesy Romaine Dennis, Good Time Classics goodtimeclassics.net)

Right: It was wise to specify the optional power steering. It cut turns lock-to-lock from 5.3 to 3.5! Overall, the feel at the helm was very Valiant. (Courtesy Romaine Dennis, Good Time Classics goodtimeclassics.net)

Far right: In August '64, *Motor Trend* noted an unwieldy 5.3 turns lock-to-lock for manual steering, but Barracuda could seat a friendly five! (Courtesy Romaine Dennis, Good Time Classics goodtimeclassics.net)

Styling & value

Critics agreed that a pony car that could double as a small family car was laudable; however, wasn't there a little too much Valiant in Barracuda? Bob McVay said *Motor Trend*'s Barracuda was mobbed wherever it went, but the general consensus was that Barracuda and Valiant were too similar. Until Mustang came along, sporty Detroit styling dictated a fastback coupe. Around this hour, even American Motors planned a snazzy fastback based on Rambler American. This Tarpon coupe never entered production but, in line with Plymouth thinking, it cost an awful lot to tool up for a unique body.

Plymouth saved time and money with Valiant Barracuda. The coupe project started after Mustang, but arrived in showrooms on April 1, 1964, 16 days before Henry's horse. Barracuda garnered 23,443 sales in that truncated model year, but Mustang was soon outselling the practical fish ten to one. Mustang was based on the humble Falcon; Ford division manager Lee Iacocca had put money into a new notchback body, which the public went ape over.

Value-wise, Barracuda made a strong case. Comparing 1965 fastbacks, Barracuda 273 V8 cost $2571, with more interior space and brake lining area (153.5in² versus 127.8in²) for its four wheel, 10in drums. That larger Chrysler senior compact body was commodious. The big glass pane added rear weight, improving traction and weight

You are observing the most popular Barracuda. In 1965, the super fish hit a sales zenith of 64,596 units. Valiant even began adopting Barracuda touches. (Courtesy Romaine Dennis, Good Time Classics goodtimeclassics.net)

distribution. The high wrap-around rear fascia improved body rigidity. With Chrysler torsion bar front suspension too, Barracuda was, in a word, roadable.

Sadly, $2697 fastback 1965 'Stang came with a 200 horse 289 V8, and *that* styling. The public was sold on the Ford's looks and optionability. Barracuda's special touches versus Valiant included: woodgrain-look plastic steering wheel, custom wheel covers ($34.15) and exclusive Metallic Gold exterior. Engines ran from the standard 170in^3 Slant-6, to a 225 big brother and 273, 180bhp, 2bbl V8. The last of these was introduced in Valiant in January 1964.

The 225-6 carried a $47 surcharge, with a $131 fee for the 273. 90 per cent of 1964½ buyers went the V8 route. Combining the 273 with the Mopar designed A-833 four-speed transmission and the optional performance rear axle, Bob McVay's *Motor Trend* Barracuda achieved wheelspin and 0-60mph in 11 seconds, the quarter-mile in 18.2 seconds at 79mph and 102mph top speed.

Commando 273 & Formula S Rallye pack

The two barrel 273 responsible for this performance made peak power at 4200rpm, 260lb/ft at 1600rpm and had an 8.8:1 CR. It came with a Ball & Ball BBD 38435 carb, but more could be had.

This 225 has a Clifford Performance intake manifold. Jack Clifford was the doyen of domestic six-cylinder power. His company slogan was 6=8! (Courtesy Romaine Dennis, Good Time Classics goodtimeclassics.net)

Car and Driver **questioned the fish's limited performance debut options. However, by 1965, Barracuda had lo po and hi po 273 V8s. (Courtesy Carrie Leetham, hanksters.com)**

Offenhauser Products did aftermarket intake manifolds of freer design, that could seat one or two 4bbl carbs. A hotter cam, higher comp pistons and free flow exhaust could take advantage of more carburetion. They had a name for an engine like that, they called it The Commando!

The Commando 273 made 235 ponies at 5200rpm, good enough for a theoretical 121mph. There was also a useful 280lb/ft at four grand. To achieve this sound revvability, Mopar magic provided a solid lifter, 10.5:1 CR motor, with dome pistons and 248-degree duration cam, versus a regular 240 degrees. Compared to the flat top piston 273, valve overlap increased from 16 to 26 degrees. Valve lift on intake and exhaust sides respectively changed from 0.395in/0.405in to 0.415in/0.425in.

Slipper-type skirts were included, with two steel inserts for the aluminum dome shaped pistons, which projected further into the wedge shaped combustion chambers. The Commando pistons were 1.37oz heavier than the stock 273 LA items, so reciprocating mass was maintained with lighter piston pins. Single became dual breaker point ignition, set at 10 degrees BTDC at idle, not five degrees. Commando 273 had a Carter AFB four-barrel, with 1.44in primaries and 1.56 secondaries.

The rumpety rump exhaust note came courtesy of a straight-through, rather than reverse flow muffler – though this was quietened down for '66 MY, due to complaints from some members of the fuzz and public. In the September 1966 issue of *CAR*, European correspondent Jerry Sloniger said its test '66 Barracuda Formula S automatic matched a four-speed Vette in a traffic light race in Sinsheim Germany. The best independent figures came from *Car Life*'s June '65 report. Zero to sixty in 8.2 seconds, the quarter-mile in 15.9 seconds at 85mph and top speed of 118mph with four-speed efficiency.

All this with 14-17mpg economy. It looked nice underhood too: oil filler, breather and crankcase ventilator caps were chromed. The rocker covers had extruded aluminum trim stuck on, and

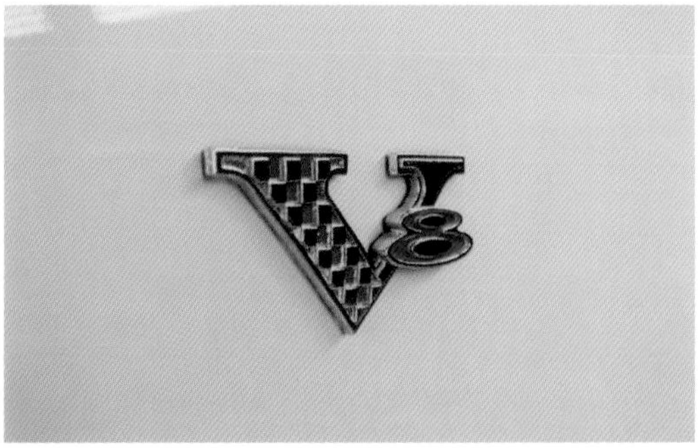

This base 273 made 180bhp at 4200rpm and 260lb/ft at 1600rpm on regular. Then, as now, you just couldn't have a pony car without a V8! (Courtesy Carrie Leetham, hanksters.com)

Pony cars were known for their often optioned, upscale interiors. The value-added lavishness pleased both public and dealers. (Courtesy Carrie Leetham, hanksters.com)

In August 1964, *Motor Trend*'s 180 horse Barracuda 273 four-speed did 0-60mph in 11 seconds. The quarter-mile came in 18.2 seconds at 79mph, with recorded top speed of 102mph. (Courtesy Carrie Leetham, hanksters.com)

a 'Commando 273' air cleaner callout decal was the performance writing on the wall. The Commando 273 could be ordered separately, but was best experienced as part of the $258 Formula S Rallye pack.

The fish that handled

This complete enthusiast package drew upon the R&D work of Chrysler's successful rally program. Richard Petty tested the prototype, dubbed 'Big Red,' at the Chelsea Proving Ground. Here, 127.3mph was recorded. For control, larger 0.87in front torsion bars joined a 0.87in front swaybar. Rearwards, six leaf springs were in residence, up from four-and-a-half. Leaf spring width had also increased from 5in to 5.5in, front spring rates from 120 to 150lb/in. Outback, the figure rose from 105 to 115lb.

There were HD shocks all around. Plymouth engineers claimed this mini rig could register under three degrees of roll in a 0.3G turn. Partial credit for this should go to the nylon bias-belted Goodyear Blue Streak tires. Chrysler-designed and Goodyear-built, they sported an industry-lowest cord angle of 29 degrees. These tires came with a protective blue coating. If not washed off, it made 'em resemble whitewalls! They measured 14x6.95in, and were fitted to Formula S's HD 14x5.5in rims.

Standard Barracuda rims were 13x5in units. Six-cylinder cars had 13x6.50in footwear, V8 coupes were on 13x7in. Formula S was something else. Its fast power steering had 15.7:1 ratio, not the 5.3 turns lock-to-lock 24:1 ratio. Fast manual steering (16:1) brought 3.6 turns, with worm and ball nut system. Power steering involved rack and sector; less precise, but less Charles Atlas muscle required! Most went for the power steering option which, as usual, had less feel than domestic rivals.

Chrysler marketing called its power steering 'Constant Control', and the Formula S pack had 'Firm Ride' shocks. With the Goodyear Blue Streaks, one could pump up the tires to 35psi for spirited maneuvers, sans harsh ride. As for brakes, four-wheel 9in drums were standard Barracuda fare in 1964. In 1965 this improved to 10in on V8 models. That year the planned availability of front disks was stated. So it was that in '66 MY, dealer-installed front disks arrived.

In the September 1966 issue of *CAR*, Jerry Sloniger said front disks were seldom Barracuda specified. The package entailed vented 11.125in disks, gripped by four-piston callipers. These Kelsey-Hayes items were joined by 10x1.7in rear drums. *Road & Track* tested the 82-buck power assisted option, in March 1966. Its Barracuda S stopped six times at 0.5g deceleration with zero fade, and no swerving. The sporty 'S' called out for the Sure-Grip lsd. This unit was available

A 1966 Barracuda Formula S in AA1 Silver Metallic. This coupe was Hamtramck Assembly (Detroit) built, and was shipped on November 19 1965 to Brooks Motors, Inc (Jacksonville, FL). (Courtesy Terry Trace)

Right: The high output Commando 273 V8 punched out 235 horse at 5200rpm. A solid lifter 10.5:1 CR motor meant serious small block GO! (Courtesy Terry Trace)

Far right: This 1966 Barracuda Formula S came with two rare options: factory-fitted a/c, and a dealer-installed disk brake package. The former has since been removed. (Courtesy Terry Trace)

with 2.93 and 3.23 rear axles alone. Mr Sloniger found Sure-Grip allowed the enthusiast to put down the same amount of power, wet or dry.

Jerry Sloniger also observed that Barracuda S both handled and stopped better than the much more expensive Mercedes 250SE coupe that *CAR* had just tested. At home, Barracuda criticism seemed centered on looks. *Car Life* felt Barracuda S's striped appearance made it look like a giant Easter egg! That decoration involved a fat band stripe over hood, roof and tail, flanked by thinner stripes on each side. *Car Life* also didn't much care for the Barracuda's sporty option wheel covers.

1964-66 revisions

Plymouth slowly distanced Barracuda from Valiant. In 1964 Valiant tailscript abounded, along with 'V' logo rear deck medallion. In 1965, just the V logo was retained. For '66 MY, Barracuda received a fish logo, which replaced the V logo. The fish symbol also lived in a front-fascia, mid-grille medallion. In Canada and other markets, the 'Valiant Barracuda' title remained until the end of it's A-Body era. Soon, Valiant was adopting Barracuda's fancy ways. For '65 MY, Valiant adopted Barracuda's fenders and taillamps, wagon aside. For their respective shared grilles, by 1966 Barracuda went egg crate, while Valiant stuck behind bars.

By 1965 the 170in^3 six was only the North American base motor on Canuck Barracudas, and factory a/c became available. 1966 saw the option of a center console, and deluxe models possessed fender-top turn-signal

This Barracuda Formula S was such an early '66 MY build, it still sported the non-center divided hood/roof/tail stripe decal style. (Courtesy Terry Trace)

By 1966, the Barracuda Formula S was critically acclaimed as a competent grand touring package. Good value at a little over three grand. (Courtesy Terry Trace)

The 1965-66 Barracuda Formula S was a high output, handling predecessor of the respected 1968 Barracuda 340-S/1969 'Cuda 340. An all purpose road car, with a practical dimension. (Marc Cranswick)

indicators with stylized fin motif. Worthy Barracuda was a solid commercial success, and reached a nameplate sales record of 64,596 in 1965.

A 1966 Formula S ad implored enthusiasts to "Take a standard eight-cylinder Barracuda and light a fire under it ..." the brand slogan was "Let yourself go ... Plymouth," and there was a Plymouth National Barracuda Club: all indicative of the burgeoning youth market, and the enthusiasts that many pony cars were bought by. 1966 represented the third year of similar Barracuda styling; bigger bumpers and squarer visage aside. The optional tach had moved from its obscured, sub-lightswitch, left of steering column position to full-size round gauge matching the speedo.

The two large gauges were on either side of the steering column, flanked by two smaller gauge pairs on the left and right. From left to right the six gauges were: coolant, gas, speedo, tach, oil pressure, and alternator. Barracuda Formula S was a rare domestic, with full instrumentation.

On the downside, the former Hurst four-speed shift linkage was now less precise. It was now from Inland Steel. It had positive reverse lockout, with 'pull on lever' detent. Whether 'S' tach or vacuum gauge, it made little difference. With expectation of a restyled Barracuda in the offing, sales dropped to 38,029. Not bad for a single bodystyle, and Camaro was coming soon.

From across the pond

One of the great rally drivers of the '60s Pat Moss, sister of Stirling Moss and wife of Saab rally ace Erik Carlsson, had success in the Barracuda. At the 1966 Tulip Rally, Pat Moss and Liz Nystrom picked up the Coupe des Dames. Best placed lady team perhaps, but the pair thought they could have taken Barracuda to outright victory.

Another British connection concerned the UK arm of Chrysler Europe, and chief designer Roy Axe. For years it had been thought that the Sunbeam Rapier was a scaled down 1964-66 Barracuda. However, in 1993 Roy Axe said, "It has often been suggested that the Rapier was influenced by the Plymouth Barracuda but I can categorically say that we didn't know anything about the Barracuda when working on the Rapier."

A coincidence of two companies going for a similar goal. Like Plymouth, the Rootes Group was constrained stylistically by having to use sedan parts from the Hillman Hunter.(1)

Barracuda 1967-69 – The stylish fish

Plymouth completely revised Barracuda for '67 MY. There were now three bodystyles, and a reskinned look that sat atop a longer 108in wheelbase. The bodystyles and wheelbase matched Mustang, and that was the whole idea. Even so, Barracuda was still A-body based.

Now, more than ever, its dimensions seemed more senior compact than compact. The bodies, beautiful, now ran from familiar fastback, to a ragtop, and Italian-looking notchback hardtop. The fastback remained the best-selling Barracuda, gaining in popularity throughout this era.

Notchback was 100lb lighter and

100 bucks cheaper than fastback. The new notchback got off to a shaky start, however: the first 2000 units possessed a distorted rear window. No recall was effected, but Plymouth said dealers would replace the pane under warranty if approached by owners.

The rear pane on fastback was smaller now, answering criticism concerning hot rear compartments. However, rear visibility was claimed to be unaffected. Dimensions for fastback were now 193/71.6/53.5in for length, width and height respectively. Length was 5in up, with front track 1.8in wider. Barracuda's engine bay was also 2in wider. Plus, all Barracudas now rode on 14in wheels. No kiddie rims that would drop you in a rut or pothole. The front wheels had moved 2in forward, aiding styling flow. This necessitated a stiffened hood and, in turn, cosmetic ribbed aluminum cover plates, to conceal the stiffening creases.

With the high wraparound rear fascia sheet metal gone, fastback now relied on a linkage bar between the rear wheelwells. Barracuda's new profile showed a startling lack of chrome ornamentation. On '67s, the leading front fender edge carried chrome Barracuda script, with fish insignia placed underneath. Formula S badging now lay behind the front wheelwell, not in front anymore. The chrome script moved behind the wheelwell on 1968 Barracudas, and back in-front on '69s.

Complying with the fashionable Coke-bottle styling look, John E Herlitz and John Sampson's new 1967 styling effort had the economy of shared front and rear bumpers. Just as well, since much of Chrysler's cash was going towards compliance with the latest federal standards: a dual circuit braking system with dash warning lamp, four seatbelts, four-way emergency flashers, energy-absorbing telescopic steering column and padded steering wheel boss,

Still on the A-body chassis, but 1967's Barracuda had all-new sheet metal styling, as evidenced by this '67 Barracuda Formula S fastback. (Courtesy Steve Sundberg)

The hi po Commando 273 was still around in the revised Barracuda line-up of 1967. Mods here include Mopar Performance Electronic Ignition, and removal of the factory a/c underhood hardware for a cleaner look. (Courtesy Steve Sundberg)

Top and right: For more flowing styling, the front wheels were moved 2in forward on the new Barracuda. The Valiant resemblance was less noticeable, compared to the 1964-66 era. (Courtesy Steve Sundberg)

As was normal for the era, a/c was a pricey option at $318.50. The Formula S pack fitted was a mere $177.50! Owner George Sundberg and pet dog Kuda are pictured. (Courtesy Steve Sundberg)

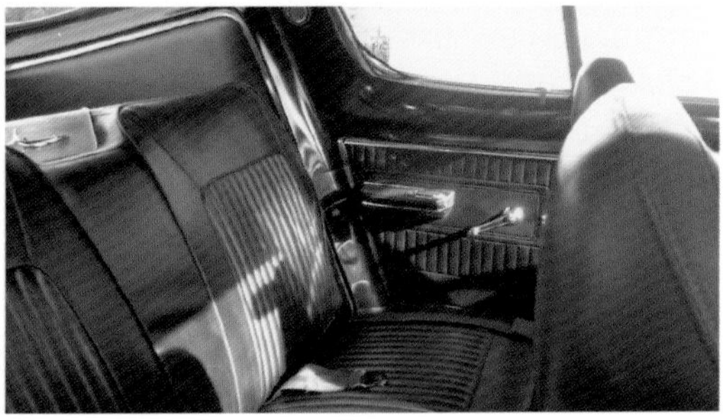

Barracuda's senior compact size implied a more usable back seat than Mustang or Camaro. This coupé had order code 328 'High' grade trim. (Courtesy Steve Sundberg)

lateral door pull handles, plus recessed safety knobs and switches.

Ralph Nader was putting the squeeze on Detroit, indeed, to make it broke at any speed! The fastback's fold-down rear seat was redesigned, with a central latch that locked the seat up or down. Plus, the gas tank filler neck was repositioned in the trunk area. Even so, 1967 marked more action for the enthusiast youth market. In ads and literature the slogan was "Plymouth is out to win you over this year ..." and was doing so with a naughty-but-nice, devil-tailed heart logo!

Revised specifications

It was business as usual with the TorqueFlite auto. Still regarded as the best domestic autobox, I6 and V8 Barracudas had the same 2.45/1.45/1.00 ratios for first, second and third respectively. Three-speed manuals had first through third ratios of 2.95/1.83/1.00 on the sixes, 3.02/1.76/1.00 on the V8. The only V8 optionable with the three-speed was the base 273. Still on 180 horses, with 8.8:1 CR.

Standard rear axle ratio was 2.93, with 3.55 and 3.91 rear ratios now available with Sure-Grip lsd. Front torsion bar diameter was 0.83 in on I6 Barracudas, 0.85in on 2bbl 273 and 0.87in with Formula S pack. Chrysler engineers wrote "none" in the six-cylinder front swaybar order blank. Standard suspension was an ordinary grocery getter setting. Ride rates were 85 and 110lb/in, front and rear respectively. Formula S cars raised that to 103 and 136lb/in, and cured the suspension of bottoming out on rough roads. The pack's 0.88in front swaybar could be separate optioned on V8s for 20 bucks.

Formula S brought decorative medallion, and six 55in long HD rear multileafs, not the standard five. The '67 Barracuda ad proclaimed, "You'll never want to go straight again;" there was even mention of "steer-by-throttle." The agile fish was that kind of car – and thoughtful too. The optional buckets and head-restraints were comfy. Barracuda came standard with split front bench, and a louvered panel near the door bottoms for stale air extraction.

Barracuda's revised dual-circuit braking system, had optional front discs sized 10.79 inches. These were gripped by four piston callipers and Raybestos Manhattan 19M brake pads. 10x1.75in drums lived out back. Normal V8 brakes were four wheel drums sized 10x2.25in front and 10x1.75in rear. Six-cylinder cars were limited to 9x2.25in drums front, and 9x2in drums at the back. The 1967 Barracuda range offered no brake proportioning valve.

Underhood excitement – Commando 273 & 383

Road & Track's '67 Barracuda, Camaro, and Mustang comparison showed the value of Mopar power trains. Despite a 330lb weight penalty over 225bhp Stang, and 70lb disadvantage compared to Camaro 327, the 235bhp Barracuda fastback out accelerated four-speed Mustang, and matched four-speed, 275 horse Camaro!

Barracuda was a half-second faster in the quarter-mile than Mustang. The Plymouth tied with Camaro 327 on 16.9 seconds, and was only 4mph slower on top speed than the Chevy. 116mph versus 120mph, not bad for a little solid lifter 273. Only some small-blocks were hi po in name and nature.

R&T observed Mopar Formula S suspension was equivalent to Mustang's GT pack. Mopar's S package now came with Firestone D70-14 Wide Oval Sport tires for HD 5.5in rims. These tires were better at channelling away water, thanks to a smaller tread block. Prototype '67 Barracudas were still on Goodyear Blue Streaks. As an alternative, BF Goodrich radials could be optioned. They promised better mileage and handling, but a harsher ride, because the suspension was still tuned for bias belted tires.

More worrying was *R&T*'s statement that they had been driving around a lot before '67 Barracuda's formal introduction, and said fastback had sparked little public response. Plymouth was hoping underhood V8 options would make the difference. Originally the 318 from Plymouth's full-size Fury was readied for Barracuda service, making the engine compartment 2in wider, and the 318's heads narrower. However, to compete with Ford's 289/390 combo and Camaro's 350/396, Plymouth had to proffer more.

That more was the 383 big block, which complemented the 273 V8 in '67 MY. As a sign of the times, voices were immediately raised, concerning the 440's absence. Being in the same family, it would slot in just like 383. Plymouth responded that would be overkill. *Car and Driver* mentioned Shelby's 428, Gurney's Cougar 427, Hurst Firebird 428 and Yenko Camaro 427. However, they weren't factory jobs, and Barracuda 383 matched Mustang 390 acceleration, which was the whole idea.

The Barracuda 383 implied two mandatory options. They were the $177.50 Formula S pack, and $69.50 front disk brakes. The Code 62 383 V8 was $52.30, an lsd $41.45, tach $48.70, buckets $32.35, Sport Stripes $30.65 and undercoating $15.40. Code 411 air-conditioning was a huge $318.50, which explained its seldom ordered nature. A/C was a pricey option on most cars of the day. Many resorted to the aftermarket, with systems like 'Comfy Kits'.

However, you couldn't get a/c with the 10:1 CR 383 motor and, more importantly, power steering was unavailable. No room, and time, for the respective compressor and pump to be placed in the engine bay – the 383's exhaust manifold was in the way. That meant standard manual steering, which was too slow, or Code 628 fast manual steering. The latter was too heavy. Muscle cars not only had muscle, they needed muscle to operate them! It seemed 'ol Henry had caught out Walter P.

The Barracuda 383 certainly had muscle, 280 horse at 4200rpm and 400lb/ft at 2400rpm. Not bad for a 3300lb bolide, but less than the usual 383 practice. More restrictive headers were necessary to snake that big 383 into little 'ol Barracuda. *Car and Driver*'s April '67 test of a Barracuda Formula S 383 notchback automatic, revealed 0-60mph in 6.6 seconds, 0-100mph in 18.7 seconds and the quarter-mile in 15.4 seconds at 92mph. The 383 came with a special Code 395 HD TorqueFlite costing $216, that made buckets and console mandatory. High output 273 and 383 V8s shared Carter 4bbl AFB primaries and secondaries. They were sized 1.44in and 1.56in respectively.

Hi Po 340 – 1968's happy medium

If the 273 was too little and the 383 too much, then Chrysler's new 340 was bound to please Goldilocks! In the 1967 Mobil Economy Run from LA to Detroit, a two barrel 273 V8 Barracuda achieved 22.3mpg. By 1968 the 273 was deep sixed, replaced by the 318 2bbl as the new starter V8. However, by the late '60s no one was interested in economy runs, or base V8s, certainly not Goldilocks. Anyway, the new Mopar 340 was possibly the best high output small-block V8, Detroit had to offer. It weighed about the same as the 273!

The 340 was a development of the 273/318 family, and had a bore larger by 0.13in compared to said 318. However, it wasn't safe to bore out a 318 like that. The 340 was underrated at 275bhp. That figure truthfully came in at 5000rpm, but the 340 could surely make more. The NHRA said the true figure was 310 horses. The advertized 275bhp was to keep a lid on insurance premiums, which were already prohibitive on pony and muscle cars. Compared to the 273's single-plane intake manifold, the 340 had a two plane affair for its 4bbl Carter AVS to sit on.

Torque was 340lb/ft at 3200rpm. As usual with a high output, small-block OHV V8, torque was sacrificed to make

Far left: A 1968 Barracuda 318 V8 automatic. Plymouth widened the Barracuda's engine compartment and narrowed the 318's heads. All to allow the full-size Fury's V8 to fit. (Courtesy Jeff Getz)

Left: Power/torque figures for Barracuda 318 were 230bhp at 4400rpm and 340lb/ft at 2400rpm. *Sports Car Graphic*'s December 1967 test of an automatic Barracuda 318 recorded a 17.9 second quarter-mile at 77.25mph. (Courtesy Jeff Getz)

top-end power, so a great amount of valve overlap made the 10.5:1 CR 340 enjoy a rev or two. And as per hi po V8s, there was a manual transmission cam, and one for slushboxes. For four-speed 340s, you couldn't have a three-speed manual, timing was 26-70 intake/78-26 exhaust with 276 degree intake duration and 284 degree exhaust duration, associated with 52 degrees of overlap. Intake and exhaust valve lift were 0.445in and 0.455in respectively.

Automatic 340s had timing-retarded four degrees on both sides. This implied intake and exhaust cam duration reduced to 268 and 276 degrees respectively. Lift for intake and exhaust valves were cut to 0.430in and 0.445in respectively. 273/318 V8s came with intake and exhaust valve diameter sizes of 1.78in and 1.50in respectively. The 340 raised those sizes to 2.02in and 1.60in respectively, outclassing Chevy 350 in the process.

One could receive all this, and more, as part of the latest Formula S Group, costing $186.30. The 340 engine, duals, HD suspension, front swaybar, 'Firm Ride'

After promising '67 sales of 62,534, Barracuda fell back to 45,412 in 1968. The pony car market had already peaked in 1967 for all players. (Courtesy Jeff Getz)

This '68 SS/A Hemi 'Cuda was a personal favorite of Ronnie Sox. Even after retirement, Mr Four-speed would drive a replica in nostalgia racing events. (Courtesy Bill Crittenden, carsandracingstuff.com)

shocks, Red-Streak Wide Oval Firestone E70-14s, 14x5.5in HD rims, and the usual Formula 'S' medallion. In 1968 only, such equipped Barracudas came with 340in^3 hood plaque callouts, decorating the faux-intake trumpet impressions. Matters could be jazzed up further with the $53.20 Sports Accessory Group. This brought sports steering wheel, custom sill molding, rally cluster (trip-o-meter, odometer, 150mph speedo), and woodgrain instrument panel.

More importantly the 340-S engine could be combined with power steering, air-conditioning and four-speed. All factory fitted. In 1968 a Barracuda 383 couldn't have those three items, but it did now have 300 horses. The Barracuda

Sox & Martin's mechanic, Jake King, built and tuned amazing Hemis to help Ronnie Sox stay competitive. (Courtesy Bill Crittenden carsandracingstuff.com)

340-S could do and be many things, in '69 'Cuda 340 form, it even defeated a super German, according to *CAR Life* magazine. Their November 1968 issue spoke of said 'Cuda besting an unnamed German high-performance car, tested by an undisclosed magazine. The car was the 1969 Mercedes 300SEL 6.3, and the magazine was *Car and Driver*.

The 'Cuda 340 beat the Mercedes 6.3, three out of three times at the drag strip. A small, initial lead off-the-line, which 'Cuda held each time. Once both cars were in top gear the 'Cuda pulled away, until the testers ran out of road! All this, with a saving of ten grand. Both cars were classed as senior compacts, with the 'Cuda 4in shorter. That was the only dimensional difference. The 'Cuda was 400lb lighter than the luxo a/c et al automatic Mercedes. With the four-speed 'Cuda having a 3.91 rear axle, the Mercedes 2.85:1, and similar power, it's little wonder 'Cuda pulled away in top.

CAR Life did note the Mercedes was geared for cruising, not sprinting. However, the 'Cuda's 120mph top speed, wasn't that far off the German's 135mph figure. Plus, the Barracuda, with small-block V8, could handle and ride reasonably. Britain's *Autocar* magazine tested a 2bbl 273 Barracuda V8 automatic in their August 31st 1967 issue.

It rode very well on Belgian pave, sans bottoming out. There was no weave or float at the UK's new 70mph limit, and it coped well with British back roads (B roads). The test also stated that the windshield wipers stayed firmly on the glass during 90mph cruising. So the Barracuda's torsion bar suspension had the right stuff.

Autocar's test Barracuda rode on Goodyear G800 radials. They were part of a single export spec, which included 2bbl 273, automatic, front power disks, Sure-Grip, HD suspension and front swaybar. With Mustang and Camaro in mind, *Autocar* said," ... ultra-smooth transmission, punchy engine and better

From left to right: Buddy Martin, Ronnie Sox, and Jake King, with their super successful '68 Hemi 'Cuda at the 1968 Spring Nationals. (Courtesy Buddy Martin)

than average adhesion our staff were unanimous in liking it more than others of its kind." Barracuda 340-S was like that, only more so.

The proof was Chrysler's rally ace Scott Harvey winning the 1968 Shell 4000 Rallye, in a 340 powered Barracuda. Drag racing legend Ronnie Sox drove a slightly modified 'Cuda 340 fastback four-speed, to a 13.33 second quarter-mile at 106.5 mph. This car used the stock 340 four-speed's Borg and Beck 10½in diameter clutch. Regular 318 cars had a 9½in equivalent. The mods were recounted by *Hot Rod*'s Steve Kelly, in the December '68 issue.

A Holley four-barrel carb, Hooker four-tube headers, recurved ignition and blocked vacuum advance, so only centrifugal action was used. Hardly any wheel hop on acceleration, or axle tramp on braking were experienced. The mods were modest, but would likely have broken emissions regs. *Motor Trend* got the best times for a stock 'Cuda 340. Zero to sixty in 6.3 seconds, and a 14.22 second quarter-mile at 99.1mph. So it was fitting that Plymouth wanted to initially call the model Mopar 340, MoPar being drag strip slang for Plymouth.

The *Motor Trend* car, tested and photographed two-and-a-half months prior to the formal '69 MY intro, had Mopar 340 callout decals on the lower front fenders, behind the wheel wells. All displayed in the October '68 issue. However, the example *CAR Life* faced off against the Mercedes 6.3, was tested two months before formal '69 MY release. This car shown in the November '68 issue, had the final choice of 'Cuda 340 nomenclature.

Either way, Barracuda 340 was a tough sell. In the October '68 *Motor Trend* issue, Eric Dahlquist said few had chosen the mighty 340 fish in 1968. Chalk this up to stiff Ford and GM competition, plus the AMC AMX 390. There were many ways to go fast, and the pony market was cooling. Dahlquist said Chrysler should have publicized Barracuda's excellent handling and load carrying properties better. As it stood, '69 'Cuda 340 was promoted as a junior Road Runner. That is, performance and HD suspension, sans trinkets and high price tag.

'Cuda 340 came with four-speed, front bench and Formula S suspension/rims. In 1968 Firestone E70-14 Red Streak Wide Ovals, were found to have traction nearly as good as slicks. Optional Goodyear Polyglas E70-14s, promised good wet weather grip, handling and ride comfort. 'Cuda 340 decoration ran to a couple of forward facing hoodscoops, plus special hood and rocker panel paint. New, attractive, 14x5.5in petal-shaped alloys were optional. So, too, were power front disks, power steering and luxo touches. Accountants and marketing were for the 'Cuda 340 set menu, and cost ruled out a ragtop version.

Charger Power 225 I6 & Hemi 426!

There is more than one way to skin a cat, there were other A-body high-performance avenues. In South Africa reliance was placed on the trusty 225 Slant-6. South Africa did like large engines, and had been known to assemble RHD versions of American cars from CKD kits, all to overcome the country's high import duties. However, by the early to mid '60s, there was more demand for smaller, economical fare.

The South African Barracuda featured a worked 225 I6. The enhancements were 2bbl carb, sports cam, 9.3:1 CR and low restriction exhaust. The result was 190 ponies, and better handling given reduced weight in the nose. In America, drag racing, NASCAR and, increasingly, SCCA Trans Am racing, held sway. For the first class, Chrysler decided to formally offer something which quarter-mile fiends and specialists had been creating themselves. Namely a 426 Hemi-powered Barracuda.

Around 50 of these 1968 Super Stock 'Cuda Hemi 426s were constructed. Plymouth sent Barracudas to Hurst for final conversion work. Hood and fenders were fiberglass, and the front bumper and doors were made of light gauge steel. Light Chemcor side-glass helped pare down the pounds. The stripped interior had a Bostrom driver seat, the 135 amp battery was moved to the trunk and the right shock tower and brake master cylinder, were also moved. All necessary to slot in the 12.5:1 CR dual Holley quad 426 Hemi, with cross ram intake manifold of $1\frac{11}{16}$ x $1\frac{11}{16}$in section diameter.

Hooker competition headers, exhaust and muffler were included, along with HD radiator and roller timing chain, unsilenced air cleaners, plus deep-groove fan-drive pulleys, to stop belts flying off in high-rpm situations. Racers could option a four-speed (Code 393) or TorqueFlite auto (Code 395). Plymouth specified a Dana 9¾in HD axle with 4.88 ratio on four-speed cars, which had special shift linkages for space clearance. Indeed, Hurst took care of the special rear axle assembly. TorqueFlite machines possessed an 8¾in differential with large stem pinion gearset, HD axle shafts and 4.86 rear axle ratio.

Cars were delivered with street tires, shallow oil pan, custom offset 15 degree rear rims, Sure-Grip lsd, and front disks with a 4½in bolt circle pattern. The BO29 Barracudas were May delivered, ready for the NHRA Spring Nationals. The nature and purpose of this Hemi powered car, similar to its Dart equivalent, were outlined in a letter from Chrysler-Plymouth general sales manager RD McLaughlin, sent to dealers and racers.

The February 20th, 1968 letter spoke of a 3000lb fastback, made in small numbers during May 1968. It directed interested parties to use the Barracuda Order Form, and tick the Body Code BO29 box. The implication was a Business Coupe interior, industry parlance for no back seat. The Hurst remote mounted floorshift had reverse lockout, and worked a 10½in HD clutch. Automatic cars possessed a high-speed stall torque converter, large drive lugs and $\frac{7}{16}$in diameter attaching screws. The TorqueFlite came with a manual shift kit.

There were many things Super Stock Barracudas didn't have: forget about body-sealer, heater, sound-deadening, silencing pads, outside mirrors, body color paint, or right-side seatbelt. It wasn't possible to order anything, and Chrysler's usual 24 month/24,000 mile warranty, five years/50,000 miles on power train, didn't apply. The advice letter warned of high idle speed, cold natured intake manifold that would cause rough running in cold weather, high oil consumption and engine noise. There would be frequent adjustment of ignition timing, autobox brake bands, along with sparkplug regapping. Ignition points would have to be adjusted and replaced more often, and engine oil would have to be changed more frequently.

A special A-pillar plant installation sticker stated, "This vehicle was not manufactured for use on public streets, roads or highways and does not conform to Motor Vehicle Safety Standards." A form letter had to be included with the

Not all racers were created equal. Ronnie Sox's skill with the four-speed stick shift earned him the title 'Mr Four-speed.' (Courtesy Bill Crittenden, carsandracingstuff.com)

In the make-what-you-race era, the 1968 Hemi 'Cuda was offered to the public as the BO29 Hemi 'Cuda. However, it was only intended for Super Stock drag racers. (Courtesy Bill Crittenden, carsandracingstuff.com)

order, signed by the customer, so that the buyer was fully informed concerning what they were letting themselves in for. One of these babies on the public highway, could wipe you out much easier than Nader's Corvair!

Flying down the quarter-mile the Mopar way

A racing legend with Hemi 426 and Chrysler A-833 four-speed experience, was Mr Four-speed himself, Ronnie Sox. It was in 1961 North Carolina that Ronnie Sox met, and raced, Buddy Martin. Martin quickly acknowledged the super proficiency of Mr Sox with a four-speed, and asked him to drive his cars. So started the incredibly successful Sox & Martin team – a drag racing mainstay outfit until 1975.

It was natural Ronnie Sox would be associated with Mopar, and the A-833 four-speed. Chrysler had the hardware that dominated drag racing, and in the '60s, four-on-the-floor was what racers used. After Max Wedge came Hemi. Ronnie Sox was at the heart of drag racing evolution. Racing an altered wheelbase Plymouth at York US 30 in 1965, he became the first driver to break the 9 second barrier in atmospheric door slammer. Sox was a Plymouth man, and the Hemi underhood featured Hilborn/Chrysler fuel-injection.

Ronnie Sox left A/FX, returning to the kind of car one could recognise, a Super Stock Plymouth with four-speed. The result were three NHRA Spring Nationals Super Stock titles between 1967 and 1970. Like Dandy Dick Landy, Ronnie Sox participated in dealership racing clinics. At Plymouth dealers he lectured on racecraft and racing. Always with the message to race on the track, not on the street.

The Sox & Martin A-body 1968 Hemi Barracuda (BO29), became the winningest car of its type in NHRA and AHRA SS/A. It was in this ride that Sox competed against Bill Grumpy Jenkins and Dyno Don Nicholson in late '60s match racing. Hurst-built Hemi Darts and Barracudas dominated the new Outlaw Super Stocks. This involved four-speed machines with over 700 horse. The category proved popular, and was adopted by the AHRA, IHRA and NHRA. The last body calling it Pro Stock.

When *CAR Life* first met the new '67 Barracuda in March 1967, they noted wheelspin and burnouts were old hat, "... now out of fashion among automotive enthusiasts." They were absolutely correct. Spinning one's wheels wasn't going to win any races on Woodward Avenue, or the official dragstrips. Your pink slip would go up in tire smoke, forcing a walk back home. Never before had the individual and corporation been so serious, about minimizing elapsed times.

Chrysler were reorganizing in the mid '60s, wishing to do racers closer to showroom stock, for Super Stock. That meant gasoline and carbs, hence the BO29 Barracuda Bandit, and its Dart Hemi counterpart. There were Chrysler sponsored racers throughout the country. Dandy Dick Landy and Ronnie Sox were prominent proponents of Dodge and Plymouth, respectively. There was no time to lose in getting those Sure-Grip lsd, 10 second quarter-mile BO29 Barracudas delivered in May, ready for the '68 NHRA Spring Nationals.

Advice was at hand too; Chrysler Performance Seminars were held at dealerships across the country, instructions given on car modification, preparation, and tuning. These seminars helped boost the sale of cars and performance parts, the latter available over-the-counter in those enlightened pre-Muskie Smog Bill days. Dick Landy's seminars, like Mr Four-speed's, were mighty popular.

Bigger than Ben Hur – 'Cuda 440

Drag racing's Super Stockers were all well and good. However, sometimes one has to pick up the groceries, or go to church. Spiritual guidance, and even divine intervention, were necessary to cope with speedy car's repayments and insurance premiums. However, Plymouth

By offering a ragtop bodystyle for the first time in the fish's life, the revised 1967-69 Barracuda matched Mustang and GM F-body. (Courtesy Wayne McDermitt)

had you covered. The 1969 'Cuda 383 had increased from 300 to 330 horses. Arch Brown thus commented on this development, in the August '84 issue of *Special Interest Autos*: "... a figure that almost defies comprehension in this day and age of small, fuel stingy 4-bangers!"

In the absolute depths that were the early to mid '80s American autoscene, when 0-50mph times were quoted, and VW Rabbit Diesels were coveted, big block Barracudas did, indeed, seem to hail from halcyon times. However, if you thought 'Cuda 383 was hot, then, in the words of President Ronald Reagan: "You ain't seen nothing yet!" By 1969, a revised steering linkage allowed 'Cuda 383 to have power steering, but Chrysler engineers had also managed to snake-in the 440 Super Commando V8!

At the time, 'Cuda 440 was the largest-engine pony car on the planet. The $345 'Cuda Group brought the 375 horse 440 V8, TorqueFlite auto, HD suspension and wheels, and Goodyear Wide Oval E70-14s. The front lower fender decal callout had been changed from 'Cuda 340 to 440 'Cuda, the displacement now resting above the name, rearwards of the front wheelwell.

The 440 supped premium gas, and had 10.1:1 CR. It made 375bhp at 4600rpm and 480lb/ft at 3200rpm, with single Carter AFB 4bbl and dual reverse flow mufflers. The supplied cam had 268-degree duration and 284-degree duration on the intake and exhaust sides, respectively. The mildness was due to the engine/gearbox combo having been lifted from sedate sedan service. The engineers went for automatic only, to protect the differential's gears. The family car origins made for slow slushbox shifting. However, there was nothing slow about 'Cuda 440.

With a *CAR Life* 0-60mph time of 5.6 seconds, and 14.01 second quarter-

This 1969 Barracuda Convertible was originally specified with a two-barrel 318 V8 and 'peg leg' 7.25in differential. (Courtesy Wayne McDermitt)

mile at 103.81mph, 'Cuda 440 turned in faster times in independent magazine testing than even the '70 Hemi 'Cuda! Credit the mere 3405lb weight for that. However, there was disappointment: *CAR Life* was hoping for the fastest time on record; the solid lifter Hemi Charger 500 did a 13.68 second pass in the '69 April issue story "The Crown Comes Back."[2]

The thing of it was, 'Cuda 440 was hard to launch. A 2000rpm brake torque start, with pause prior to nailing the gas pedal, proving to be the best method.

CAR Life called 'Cuda 440 a disturbing automobile, in the sense that it placed all its eggs in the quarter-mile basket, and was much worse in handling and braking compared to the all-purpose 'Cuda 340. However, 'Cuda 440 did work for certain buyers. It was very good for highway driving, with an untemperamental engine that delivered tremendous straight-line acceleration. Outside of parking lots and mountain road driving, it had adequate handling and braking.

Where off-the-line traction wasn't an issue, 'Cuda 440 drove home its advantage over 'Cuda 340, and just about any other car, at any price. Comparing *CAR Life* tests it cut 'Cuda 340's 30-70mph passing time from 6.2 seconds, to just 4.3 seconds! A party trick that could be a life saver, if one had to overtake an 18 wheeler on two lane blacktop. 'Cuda 440 would be great for the owner needing a reliable daily driver, that could be a strong contender, in production car related drag racing categories. The racer you could drive to the strip, and back home again – a gentleman's racer.

At this hour, one couldn't underestimate the street cred of that '440 'Cuda' decal. No wonder Plymouth put the 440 number first; it declared 'Cuda's top dog nature in a world of smaller engine ponies. An adjacent item was also noticeable – exterior lighting. 1967 Barracudas had no side markers or reflectors. '68s carried round side markers, but no reflectors, and '69s had just rectangular side marker reflectors.

The late '60s were a mix of dreams and reality. Coping with new pollution and safety regs, but still a time of rising horsepower, experimentation and customization. In 1967, Plymouth designers created the Formula SX Barracuda. This styling proposal had a fiberglass body for the latest 108in wheelbase. With 192.8in length, and 51in height, it was 3in shorter than a '67 Barracuda. The grille, headlamps and turn-signals were recessed behind a frame like bumper, for protection.

As a '60s global player, Chrysler Corp exported all three Barracuda bodystyles to Britain. Its factory-designed, right-hand drive conversion was something Camaro and Mustang didn't have. UK Barracuda convertibles and fastbacks were priced the same. (Courtesy Wayne McDermitt)

1969 was the final year for the Barracuda convertible. Similarly, there would be no GM F-body ragtops for '70 MY. Rollover safety fears, increasingly popular a/c, and changing consumer preferences, saw convertibles decline. (Courtesy Wayne McDermitt)

George Barris' SXB Barracuda

In the limited production customization field, Chrysler/Plymouth endorsed a George Barris prototype, dubbed SXB Barracuda. Based on a new 1967 Formula S Barracuda 383, the creation had power brakes and, unusually for this year of Barracuda, power steering and four-speed.(3)

However, with a customizer car it's the cosmetics that count. SXB was painted pearlescent blue and candy blue. There was a flat-black rear fascia, with patriotic red, white, and blue encircling tail stripes. A rear spoiler, and non-functional side pipes, which could be made functional, were not the only body mods – flamethrower headlights sat in a flat-black, modified grille, and parking lights were moved to the lower valence.

Such lightly modified packages were car-dealer doable. Barris did such packages for various brands of pony car. It satisfied the public's desire for a turn-key custom car, and was also a showroom traffic magnet for dealers. Plymouth worked in some mild changes during the 1968 to 1969 model year transition, such as a new front grille and silver rear fascia panel. The padded under-dash safety lip of 1968 was gone, and a *Hot Rod* prophesy would come to fruition.

In the December 1968 issue of *Hot Rod*, Steve Kelly noticed that Mustang fastback's trunk access panel was deleted at the end of '68 MY, but hardly anyone else noticed. He feared the same for Barracuda, and the fear was justified. Not only did Plymouth jettison the access panel, it deep-sixed the fastback altogether! In that roller-coaster, pony car sales ride, the new Barracuda reached 62,534 sales in 1967's honeymoon debut, but fell back to 45,412 and 31,987 in the next two years. Increasing competition and a deflating pony car market were the culprits, but Barracuda would return.

2

Barracuda & Challenger 1970-74 – The E-bodies

The fast fish becomes the ultimate pony

The 1970 Barracuda was an all-new car in a way domestic rides seldom were. It not only looked different, but was different at heart. Still residing on a 108in wheelbase, but no longer associated with the A-body. The new E-body was a truncated version of the B-body associated with intermediate Charger. So the new '70 Barracuda was shorter, but much wider. Dimensions were 186.7/74.9/50.9in for length, width, and height respectively on a Plymouth 'Cuda, that abbreviated moniker now being the official term for the sporty version.

All Barracudas now looked extremely sporty. Plymouth had abandoned the fastback, and any sense of practicality. In ad copy, the marque even boasted that the new fish had "the smallest trunk in the industry." Pony and muscle cars were all about looks and power. A utility-giving fold-down back seat? Forget about it! New bodystyles were limited to a notchback hardtop and ragtop. The looks came courtesy of John E Herlitz. This time there was no question mark over Barracuda's looks – it was universally admired.

This matching numbers '70 Plymouth 'Cuda 440-6 four-speed with A34 Super Track Pak, shaker hood and Go-Wing factory options is one hot super fish. It doesn't need any wasabi! (Courtesy John Cross)

The front and rear fascias had ovoid themes – at the front, a deep set face with single circular headlamps. The front turn-signals were placed under the hood's leading edge; well concealed, except when required. 'Cudas came with a round pair of inboard front fogs, under the front bumper. Stock chrome bumpers could have foam-backed, color-keyed urethane covers. This latter item was optional, looked very integrated, and was color-keyed to match the exterior paint.

Rear taillamps had two horizontal ribs. 340/383 V8s, or better, implied dual exhausts exiting through the rear valance. Inside, full Rallye instrumentation was possible. From left to right: 150mph speedo, offset 8k tach, combination gauge – with oil temp at nine o'clock, fuel at noon, coolant at three o'clock, alternator at six o'clock – and oil pressure. There was the option of six-way manually adjustable driver's seat, giving Chrysler the ergonomic edge over AMC and Javelin.

1969's Mod Top vinyl option, with associated floral motif, carried over to 1970. With Mod Top, one could pretend to be a hip, young *Mod Squad* character. The package had matching seat material and door inserts. A real taste of luxury was afforded by the $170 Gran package for 1970 hardtop and ragtop.

The Gran trim level included real quality-leather buckets with six-way driver-side manual adjustment, faux wood trim and overhead consolette. This last innovative item carried warning lights for 'door ajar,' 'low fuel,' and 'fasten seatbelt.' The Gran Coupe name originated from full-size Fury. GM F-body wouldn't offer a reclining seat option until 1982. BMW chimed in the same year, with overhead console warning lights. Mopar's fabulous fish was really swimming ahead of the shoal!

On a less positive note, general interior material quality left something to be desired. Even Mopar fans concede that interior plastics and door panel trim aren't of a luxury standard. Doors also clanged shut, rather than 'thunk' as with big Chryslers of old. This was partly due to the wide, frameless doors of such pony cars. Independent testing said that E-bodies were well made, in spite of this body type. One standard E-body feature was abundant passive safety.

Motivated by fear over possible fed regs, including roof-strength rollover tests, the E-body was a solid thing of beauty: there was a box-section sheetmetal rollbar under the roof, and fully-boxed roof rails, rear window and windshield headers; there were two-piece, high yield-strength, steel door beams; doors and surrounding frame were fortified. There were also the federally-expected collapsible steering column, padded dash, and steering wheel spokes.

A problem with E-body, and even imported cars like Volvo 140/160, was the continued presence of the gas tank filler neck in the trunk area. For active safety, E-body had front and rear tracks increased by 2in and 5in, respectively, versus A-body Barracuda. The fish was still a handler. In Chrysler code parlance, there was base BH Barracuda, BP Gran Coupe, and high-performance BS 'Cuda. A convertible was available for all three, if one wished, and Gran trim could be 'Cuda optioned.

For the budget minded, there was the A93 Barracuda Coupe. This variant ran from mid '70 MY, to the end of '71 MY. It was the only fish with regular fuel 125 horse 198in^3 Slant-6. Apart from a very plain interior, it also featured fixed rear-quarter glass. The normal base Barracuda and Gran Coupe started with 145bhp Slant-6 motorvation, although most optioned the base 230 horse 2bbl 318 V8. However, this wasn't what memories were made of ...

'Cuda – the sporty fish

Unlike some pony and muscle car rivals, Plymouth clearly told prospects which version was the sports model. It was plainly denoted 'Cuda. With 'Cuda, one received the tuned, sporty, domestic car that was on the rise. Not just straight line performance – handling and cosmetics were in the mix too. That last quality would assume increasing importance as 'Cuda, Boss, Z/28, Trans Am, and AMX quickly became suburban status symbols, not just performance icons.

'Cuda's base price was a competitive $3164. For such outlay one got effective, front under-bumper, inboard, circular foglights. The rear decklid spoiler, and rear fender silhouette stripes with engine size callout, were delete options. Buyers could get a credit for them. Front hoodscoops were not functional.

The original engine, gearbox, rims etc, have been stored safely. It's possible to return this 'Cuda to standard. Modifications include a 493in^3 high-quench 10.25:1 CR motor. The V8 has a 4.15in stroke, lightweight reciprocating assembly, with ported 345 CFM intake Indy EZ heads, plus 0.620in lift mechanical roller cam. (Courtesy John Cross)

As part of the 700 horse plus V8, three very modified 500 CFM Holleys sit on a heavily ported aluminum intake manifold. Fuel supply comes from a 28 gallon stainless steel fuel tank, care of a two-speed electric pump. Exhaust gases pass through a full 3in TTI system, with x-pipe crossover and electric exhaust cutouts. (Courtesy John Cross)

All 'Cudas came with HD brakes. This involved the B11 order code, and upgraded versions of the four wheel, 11in drums. A 0.94in front swaybar was also standard. 'Cuda life also implied 0.90in front torsion bars, instead of the 0.88in items of normal Barracuda. 15x7in Rallye steel rims, with Goodyear Polyglas GT E60-15s, came along for the ride.

Ride rates for 'Cuda regular were 111lb/in front and 132lb/in rear. This all came from the standard S13 HD suspension. A key stock item was the 4bbl 383 big block V8 – it made 335 horse at 5200rpm, and 425lb/ft at 3400rpm. When *Car and Driver* tested Barracuda Formula S in June 1966, it loved the car's small block V8 balance. The journal added that 383 or Hemi would wreck this quality, and that it would oppose such a move. E-body's wide nature, and 125lb weight gain over A-body Barracuda, made big block V8s a refined addition. Indeed, Barracuda was now engineered to accept this motor with power steering and a/c.

The 383 had a reputation as Chrysler's nicest street engine. 'Cuda's standard tach kept one informed of 383's activity, and the standard trip odometer showed how far 'Cuda 383 was going on a gallon of gas: 16.2mpg at 65-70mph, according to *Road Test* in June 1970. This affordable big block commenced in 1959. Master engineer Otto Winkleman was behind getting Chrysler Corp off the expensive Hemi, as the sole high-performance route. The police force certainly loved 383 V8.

Chrysler's engineers were very partial to the regular fuel 2bbl 383, which only came with automatic transmission. It made peak power and torque figures 200rpm below the hi po 383, and drank 87 octane gas. The 'Cuda 383 version had 10.5:1 CR, 284 degree duration cam, rather than the 2bbl 383's 260 degree equivalent, cam surge damped valve springs, and stronger coil. The normal family car distributor was retained.

As a no-cost option, 'Cuda buyers could try the hi po, 275 horse, 340-4 small block. Chrysler's little Mopar 'motor-that-could' bamboozled insurance companies, and kicked understeer to the curb. The entry level 'Cuda had the D13 floor shift HD three-speed box. 383 four-barrel was the first level combinable with the Hurst linkaged four-speed. 340 V8 could have the well known A-833 four-speed too.

The optional TorqueFlite 'Slick-Shift' auto had a safety button to stop overshooting into neutral. In the 340/383 class, a HD 8¾in ring gear was provided. A nice combo was 'Cuda 383 with shaker fresh-air package and a/c. The sportiest rear axle ratio for this ride was 3.23:1.

The 1970-71 shaker hood (N96) had a forward facing dual nostril molding, and there was a dashboard air

Sporting a tad over 62,000 miles, the usual A833 four-speed has been replaced by a Passon overdrive five-speed. This manual utilizes a twin disc McLeod Street twin clutch. It connects to a chrome moly driveshaft that turns a B-body Dana 60 with 3.73 gears. (Courtesy John Cross)

An inset spring hanger kit allows the Dana 60 to be mounted, so that huge 325/50-15 rear Nitto 555 Drag Radials can be used sans inner-panel clearance work. (Courtesy John Cross)

inlet control for said scoop. The same Elastomeric hood scoop material was also used for the optional, color-keyed, foam-backed bumper beauty covers; just order the front bumper (A21), or get both (A22). E-body permitted selective option ordering to a degree that even Mustang no longer offered.

In the normal Barracuda range, the 330 horse 4bbl 383 was the option box zenith. However, in 'Cuda land, one could choose the 440 Super Commando, the 440-6 or the fabled 426 Hemi. This first 4bbl motor was the 375 horse unit seen in the '69 'Cuda 440, and was the highest motor teamable with a/c; the upper two couldn't have a/c, with the 440-6 involving a triple deuce Holley arrangement. The Six Pak had 10.5:1 CR, and made 390bhp at 4700rpm, plus 490lb/ft at 3200rpm.

The trio of two-barrels could be ordered over the parts-counter of certain dealers. Stick shift cars needed kit No 3412099; automatics, No 3412100. The front Holley was part No 3412050, middle was No 3412052, and rear carb, No 3412051. When cruising, only the center carb functioned – the other two came on with the vacuum effect of a tromped throttle. The dual point distributor associated with the 383 and 440s, were represented by part numbers 2095065 and 2444366 respectively. A baffled oil pan for 383/440/426 was part No 2863983, with HD Ramcharger oil pump kit for such big blocks being part No 3412031.

A restaging of *Car Life*'s November '68 'Cuda versus Mercedes face-off. 'Cuda won this encounter, too! (Courtesy John Cross)

Only 652 Hemi 'Cudas were produced in 1970. Convertibles were even rarer, at just 14 units! (Courtesy Sam Cameron, sanfranciscosportscars.com)

The immortal Hemi 426, now in hydraulic lifter form for easy '70s living. However, one needed premium gas (99 octane or better) for the 10.25:1 CR and 425 horses. (Courtesy Sam Cameron, sanfranciscosportscars.com)

Hello, Hemi!

Hemi 'Cuda – its reputation preceded it: the Big Daddy of the engine line up. No wonder Don 'Big Daddy' Garlits built 'em up for his Darts. As a predecessor to funny cars, there was the factory drag racer Hemi 'Cuda, as successfully raced by 'Mongoose' McEwen in 1964-65; after all, mongeese do eat cobras!

Still, with a roll cage and center seat, you weren't likely to see one at the mini mall. Even subsequent fiberglass replicas were pretty exclusive. However, Hemi 426 did become available in the Dodge Coronet, Charger and Plymouth Satellite/Road Runner. There was the 1968 BO29 Hemi 'Cuda. Engine insertion necessitated this A-body having its brake master cylinder shifted forward and left. The right front shock tower was also moved.

In 1970, one could buy a 'normal' Hemi 'Cuda that was truly civilized. For the first time ever, the Hemi 426 had gone hydraulic lifter; not the cantankerous solid lifter V8 of old. Like the aforementioned 440-6, the 1970 Hemi 426 was quite docile. If seriously anxious, the Hemi could be converted back to solid lifter format, but there was no point – on the oversize Goodyear Polyglas GT F60-15s, it was a real devil getting the power down.

Also helping to get the Hemi 'Cuda off the line was Barracuda's level three S15 suspension. This 'extra HD' suspension was on 440/Hemi cars. Unlike with 340 and 383 'Cudas, the rear 0.75 swaybar couldn't be optioned, because things were very sled-like stiff already. S15 included S25 Firm Ride Shock Absorbers, which could be separate-optioned on 'Cuda 340 and 383s.

E-Body Barracuda was shorter than its A-body predecessor, causing worse weight distribution. With Hemi 'Cuda it was 59.3/40.7 per cent, front to rear. The solution? Power oversteer! (Courtesy Sam Cameron, sanfranciscosportscars.com)

Hemi 'Cuda, like 440 Six Pak cars, guaranteed a standard four-speed. However, both hot performers couldn't have cold factory a/c. They could have luxo Gran Coupe trim for $170. (Courtesy Sam Cameron, sanfranciscosportscars.com)

With level three, front torsion bars were 0.92in diameter. Front ride rate increased to 118lb/in, with rear rate at 150lb/in. Indeed, 440 and Hemi rides had 5½ leaf springs, not the usual 4½ leaf springs.

The asymmetric shocks and extra HD suspension made for flat handling, no bottoming out, and ride comfort alien to Imperial devotees. For serious drag strip fiends, Chrysler offered axle packages, which had included all the good stuff under one roof. 'Cuda 340s and 383s came with HD 8¾in ring gear; on 440-6 and Hemi cars, a Spicer built Dana 60 differential with 9¾in ring gear was standard on four-speeds.

440-6 and Hemi 'Cudas came with standard four-speed, or specially calibrated TorqueFlite. A 4.10 rear axle ratio was optional. The TorqueFlite's governor would keep the engine spinning high up the rpm scale post shift, so that the V8 stayed in its power band. For such serious quarter-mile times, there was the 'Super Track Pak' (A34) for 440 and Hemi 'Cudas. This drag strip option included 9¾in Dana 60 with 4.10 rear gears, Sure-Grip lsd, HD radiator and shroud, seven-blade fan, dual breaker distributor, and 11in power disk front brakes.

Disk brakes were not standard on any 'Cuda, and the A31 and A36 performance axle packages ruled out a/c for 'Cuda 340 and 383. Hemi 426 did provide cross bolted main bearings, HD castings and reinforced web journals. With 10.25:1 CR, outputs were 425bhp at five grand and 490lb/ft at four grand. Dual Carter 4bbl AFBs supplied premium juice for the mighty Hemi.

In reality, axle tramp put a bigger limit on braking, regardless of whether drums or disks were present, and the powerful cars really wanted cheater slicks. *Road Test*'s June 1970 'Cuda 383 report yielded a 14.4 second quarter-mile at 98.97mph. *Motor Trend*'s May 1970 article, on an automatic Hemi 'Cuda brought a 14 second pass at 102mph. The 'Cuda 383 had a locked 3.91 differential, whereas *Motor Trend*'s Hemi had a 3.55:1 rear end. Whatever 'Cuda combination chosen, it could be painted in a choice of 18 'High Impact' exterior colors.

In the pony/muscle car era it was normal for domestic cars to have special paint names, and 'Cuda's 1970 palette was pure pleasure. In the mix were Tor-Red, Lemon Twist, Black Velvet, Lime Light, Vitamin C, In-Violet, Curious Yellow, Bahama Yellow, Sassy Grass, and Moulin Rouge. One could even have Sheep Dog White! The sad thing was, just a few years later, once in the pollution/economy malaise era, all this frivolity vanished. In 1970, all one needed was a little hindsight.

The fish goes racing

In 1970 the 440-6 motor was a $250 'Cuda option, whereas Hemi retailed

What's in a name? Enough to turn a 1970 five grand car into a million dollar collector dream machine! Hemis and ragtops were built only at Hamtramck. (Courtesy Sam Cameron, sanfranciscosportscars.com)

for $871. At the time old Hemi looked pricey, next to the nearly as quick Six Pack. However, hindsight would have told the 1970 prospect that such optioning would prove to be the difference between a five figure collector car and a six figure one. In 1970, swift 'Cuda 383 or 340-4 had no surcharge, explaining their popularity.

Unless you were a racer – then Hemi motorvation became a necessity, not a luxury. With the E-body Hemi 'Cuda, Ronnie Sox continued his winning ways in 1970-72 ProStock. The car had the tuning expertise of mechanic Jake King. Ronnie Sox made the first series of hoodscoops for the dominant 1970 Hemi 'Cuda he campaigned. Keystone made the wheels for this car and the previous Sox & Martin 1968 Hemi Barracuda.

The Sox & Martin team with its E-body Hemi Cuda updated to '71 look. The team ended its formal association with Chrysler Corp in mid 1974. (Courtesy Buddy Martin)

Ronnie Sox used a '72 'Cuda in the street eliminator class, and continued with Plymouth's four-speed machines into the Lenco shift era.

Drag racing and oval racing tended to be regional. SCCA Trans Am racing was national, increasingly popular by the late '60s, and something Chrysler wanted to get into. Chevrolet, Ford and AMC were there. Pontiac was too, thanks to a Canadian Firebird Chevy small block V8 order blank. Chrysler were in need of a suitable V8. The classes involved limits of two liters and five liters (305in^3). The smaller category involved imports, the larger for pony cars. Fortunately the SCCA rulebook changed, allowing an existing V8 to be brought under the 305 cube limit.

For Chrysler, that meant destroking the 340 to 304in^3 from 3.31 to 2.96in. Like-rivals power was in the 430-450bhp range by 8500rpm, according to Plymouth engine man Pete Hutchinson. The motor started coming on from four grand. Joined to the 304 V8, was a special four-speed. Hutchinson saying many hours were spent building slick, quick, and light action four-speeds, with three-plate clutch and Hurst shift. The rear axle ratio was 4.10:1, power went to 26in wide rear tires, and four-wheel Kelsey-Hayes disk brakes arrested progress. The light V8 was set well back in the engine bay, giving excellent 53/47 per cent front-to-rear weight distribution.

Sentiment was that Chrysler had the best racing chassis and motor. In an era when you raced what you made, a homologation road car was expected. This was the '70 MY only AAR 'Cuda 340, introduced mid model year. A car which, along with its Dodge sibling, *Classic & Sports Car*'s Julian Balme referred to as one of America's last great cars of the pony/muscle car era. In driving this triple-deuce-fed bolide, Balme said "There's a momentary stutter before the other two carbs kick in and throw you hard into the next county."[(4)]

Don Mathews of *Sports Car Graphic*,

One of the 2600 AAR 'Cuda 340s built to homologate the 'All American Racers' that Dan Gurney and Swede Savage campaigned in the 1970 SCCA Trans Am series. (Courtesy Sam Cameron, sanfranciscosportscars.com)

was moved by this Super 'Cuda. In the June 1970 issue, he invented a new category for it. The AAR 'Cuda 340 wasn't a pony car or muscle car, it was an "Impulse Car." It was certainly fleet of foot.

According to *Car and Driver* the showroom car did 0-60 mph in 5.8 seconds, with a 14.3 second quarter-mile. Not quite as potent as the racer, but 290 ponies were quoted at five grand. Given the 6500rpm red line, figure a genuine 300-plus horses. Torque was 345lb/ft at 3400rpm. The SCCA rulebook said to retain factory block and heads. The blueprinted 340 block had extra material around main bearing webs, four-bolt mains and relocated pushrod holes for Hemi valvegear.

The 'Six Pack' implied triple deuce Holleys, on an Edelbrock intake manifold. Underhood sights were a picture. A large orange air filter housing, with '340 SIX PACK' callout and checkered flag decal. There was a rubber surround to the 'Cuda's NACA hood duct, and racing pins to secure said fiberglass hood. The front and rear spoilers were certainly functional. At 100mph the front unit generated 440lb of downforce, the rear spoiler produced 125lb. All to keep AAR 'Cuda 340 glued to the road or track.

The HD suspension brought a rear swaybar. It also involved rear ride height raised by 1.75in. This was to clear the rear leaf springs from the dual, reverse flow side exhaust system. The front Goodyear Polyglas GTs were E60-15s, the rear G60-15s; all wrapping oversize 15x7in rims. Further visual distinction lay with exterior graphics and paint. A black, profile-length strobe stripe ending in the word "CUDA," with red, white, and blue decoration. Mighty patriotic pilgrim, but then it would be, given AAR stood for Dan Gurney's 'All American Racers!'

The hoodscoop, and indeed the entire car, was functional. Cowl, fender tops, grille, rear fascia and rear spoiler were flat black. Aforementioned

The AAR 'Cuda 340's Hurst-shifted four-speed had shorter second and third ratios, versus the usual A-833 box. Out back was a 3.55 rear axle, with 3.91:1 a no-cost option. Three 2bbl Holleys fed premium leaded gas to the 10.5:1 hydraulic lifter, blueprinted 340. An underquoted 290 horses at five grand was the result. (Courtesy Sam Cameron, sanfranciscosportscars.com)

Car and Driver was able to get a 5.8 second 0-60mph time out of AAR 'Cuda 340. The same time _Motor Trend_ achieved in May 1970, with its automatic Hemi 'Cuda. (Courtesy Sam Cameron, sanfranciscosportscars.com)

hoodscoop (NACA duct) and plastic hood, were the same hue, too. You would expect AAR 'Cuda 340 to be a handler ... and it was. Weight distribution was 57/43 per cent front to rear. Not as good as the track car, but better than big block 'Cudas. Steering was special and quick, but the race car had zero caster. This made it ultra-zappy, but wouldn't self-center after a turn. So only very skilled drivers need apply.

The car used a special close ratio version of the A-833 four-speed. Second and third gears were shorter at 1.91 and 1.39 ratios respectively. First and top stayed at 2.41 and direct 1:1. Brakes were normal optional 'Cuda disk/drum with power assist. These good brakes, elastomeric color-keyed bumpers, racing side-mirrors, Rallye instrument cluster, and power steering were all options not included in AAR 'Cuda's $3966 base price.

In this era of ordered build sheet performance, AAR 'Cuda, apart from engine blueprinting, could be created by judicious use of the order booklet and a helpful dealer. Thus, the fast power steering involved steering gear chuck (part No 2537749) and steering link gear arm package (part No 2948713). For the 340-6 the block was part No 3577229, the heads were part No 3577050 and camshaft part No 2899206. AAR 'Cuda's front spoiler involved a right side (part No 3443352) and left side (part No 3443353) and needed 5⁄15in bolts.

The AAR ducktail rear spoiler was part No 3570369, with associated bracket and four studs coming under part No 3570359. Die-cast model maker Hot Wheels supplied sponsorship money, for the All American Racers Plymouth team. Chrysler Corp gave an additional $1 million. Unfortunately, it seemed that ordering victory was much harder. The Plymouth AAR 'Cuda 340 was fast. Dan Gurney's protégé and number two driver, Swede Savage, got three pole positions. Engine longevity, however, wasn't too good.

Dan Gurney raced his famous No 48, with Swede Savage in the No 42 AAR 'Cuda 340. In the final analysis Ford won the championship on 72 points, with Plymouth on 15 points. In 1970 it was Parnelli Jones' Mustang versus Mark Donohue in the Penske Javelin. It was unrealistic to expect Chrysler to walk into the series and take the championship

The AAR 'Cuda 340 was only available in 1970. However, using the 1971 order blanks, and a helpful dealer, a replica could be created by ordering the correct parts and packages. Only certain dealers offered such Trans Am series parts. (Courtesy Sam Cameron, sanfranciscosportscars.com)

Above: For 'Cuda's factory-backed assault on the Trans Am series there was 'All American Racer' Dan Gurney. He raced the homologated 450 horse AAR 'Cuda Six Pack! (Courtesy Dan Gurney, allamericanracers.com)

Left: Dan Gurney's team-mate, Swede Savage, achieved an E-body Trans Am best result of second at Elkhart Lake. Sam Posey was third in the Sub-Lime No 77 Challenger. (Courtesy Dan Gurney, allamericanracers.com) first time out. Ford, Chevrolet, Pontiac, and AMC had been there for years. Sadly, there would be no second chance, as Chrysler Corp Series' costs were exploding, and the major players were moving on. The industry as a whole had more pressing issues; as far as domestic pony cars were concerned, the Trans Am series had run its road course.

1971-1974 – The fish in challenging times

There were two great results for Plymouth in 1970. The new Barracuda was a success, hitting 55,499 annual sales. It helped push Plymouth to third in the sales race for the first time since 1959! 1971 would prove to be a mixed year; good for enthusiasts, not so good for business. With tightening emissions law, and low-lead/no-lead gas looming, 1971-72 was the time of the compression ratio drop, GM using its muscle to make this the quick fix in the wake of the 1970 Muskie Smog Bill. Even so, Chrysler, like Ford, alongside some imports, held onto their high comp ratios in 1971.

The mighty 440-6 and Hemi 426 kept their comp ratios, the Hemi even having identical power and torque figures. The 440-6 was down five gross horses, to 385bhp. The 440-4 Super Commando, the highest a/c friendly motor, was now a special order delicacy. So too the AAR 'Cuda 340-6 in some ways. Plymouth had hoped to continue it into 1971. Ford thought the same for Boss Mustang 302, but reality bit. However, apart from blueprinting, all the AAR 'Cuda stuff was still orderable from selected dealers. No more factory ready racer, but you could make your own.

A 1973 'Cuda 340 – originally FE5 Rallye Red with white side stripe, it was repainted TX9 black. A black interior was added as part of a rotisserie restoration. (Courtesy Ivan Goldenberg, youtu.be/tk3YgEGX3yY)

As ever, one contacted the Chrysler Performance Parts Service (6565 E. 8 Mile Road, Warren Michigan, 48091) to get a list of helpful dealers. In those specialized times, not all Plymouth or Dodge dealers could service and supply parts for Hemi cars. 1971 would prove to be the final year for the classic Hemi, and Barracuda ragtops. For a super rare collectible, try a 1971 Hemi 'Cuda convertible; only 13 were made, and just seven were sold in America!

One could still rely on the 'Cuda 340. All the figures were the same with 10.5:1 CR, 275bhp and 340lb/ft, but there were changes for '71: the 340 now used log type exhaust manifolds, not free-flowing headers, and there was a carb change from Carter AVS to Carter TQ. To beat emissions and maintain performance, primaries were smaller and secondaries were larger. The Thermo Quad involved 4bbl sizes 1.38in/2.25in, versus the AVS' 1.44in/1.69in. The 340 continued its great popularity. Although over 200lb heftier than Duster / Demon, 'Cuda 340 was generally swifter in the quarter-mile. It seemed the E-body chassis allowed 'Cuda to hook up better than the A-bodies.

The normal Barracuda 383 4bbl dropped from 330 to 300bhp, and the 2bbl edition fell to a 340 matching 275bhp. Of more concern was Barracuda's 1971 restyle. The front fascia now had four circular lamps, and a six D-shaped segment pearlescent effect mid grille trim piece. The latter would be color-keyed with the optional Elastomeric bumpers. For the rear fascia there were

The hi po 340 was Chrysler's avenue into SCCA Trans Am racing in 1970. Carrying an underquoted 275 gross bhp to fool insurance companies, it became the standard 'Cuda motor in 1972-73. (Courtesy Ivan Goldenberg, youtu.be/tk3YgEGX3yY)

now four horizontal chrome strips for the taillights, and new inboard vertical back-up light segments. 'Cuda's optional profile decal, sans engine size callout, was now a blackout item. It covered the whole rear fender and first third of the front door.

For 1971 only, 'Cuda could have a quartet of front fender gills. Six-way manual adjustment driver's seat, rear window venetian blinds, and front disks were still options; however, that last item was now mandatory on 990 CFM 'Cuda 440-6 and Hemi 'Cuda. Newly optional was a stereo cassette deck, available with or without recording microphone. 'Cuda may have had gills, but Barracuda sales were sadly all at sea: they fell to just 18,690, only slightly due to the poorly received front end restyle.

The fish gets rationed

The pony car segment hit its peak in 1967 with 752,821 sold. By 1971 that number had shrunk to half the size. Competition from GM and Ford was stiff, and a new Javelin debuted in '71 MY. So Plymouth pruned Barracuda for '72 MY. There were now no big blocks, no ragtops, nor Gran Coupes. Gone too were power windows and leather seats. The three remaining engines were Slant-6 (198/225), and the 318 and 340 V8s. The first got induction hardened valve seats, for the transition to unleaded gas. The last engine was now 'Cuda's standard and only powerplant. The 340 V8 received transistorized ignition, with 6400rpm rev limiter; it was a driveability aid addition to 'Cuda life.

The engine line up endured the industry wide compression ratio drop, and change to net power ratings. The honest, regular gas drinking powerplants saw a 100bhp 198 I6, 110bhp 225 I6, and 150/240 horses for the respective 318/340 V8s. The 340 now made 240bhp at 4800rpm and 290lb/ft at 3600rpm. Redlined at 6500rpm, this now 8.5:1 CR edition sported 360 V8 heads. Intake passages were smaller and poorer flowing. Worst of all was the first reduction in intake valve size since 1968. The 340 went from the original 2.02in to 1.88in.

Muscle and pony car fans love such exterior decoration today. However, back in the day, many invoked the delete option on the 5in high engine size callout and stripes for a street-sleeper look. (Courtesy Ivan Goldenberg, youtu.be/tk3YgEGX3yY)

By 1972, Barracuda had reverted to its popular, two-headlamp front fascia. This 1973 'Cuda 340 has been retrofitted with 1970 chrome bumpers. (Courtesy Ivan Goldenberg, youtu.be/tk3YgEGX3yY)

It seemed to cost 'Cuda one second in the quarter-mile. It was now a mid-15s coupe. The standard 'Cuda Rallye suspension was the S13 package of old, with the 0.75in rear swaybar included. However, footwear was now 14x5.5in rims wearing Goodyear F70-14s. A performance axle package involved 3.55 ratio and Sure-Grip lsd, and cost $61.45. The 340 engine cost $284.05, and could be separate optioned on ordinary Barracuda. The four-speed retailed at $198.10, and was reserved for 'Cuda. A base Barracuda I6 priced out at $2867.

Barracuda's 1972 restyle was well received. The front fascia was back to just two round lamps. The segmented grille of '71 was replaced by a central pointed, slotted aero nose post. This item clearly influenced the 1973 Pontiac Grand Am's face. In 1972-74, Barracuda's visage looked pretty much like that of the '70 MY. The tail received four equal-sized circular lamps. The optional profile stripe was redone for 1972-74, with a solid-color, upper-body third decal. The one-piece decal had a single pinstripe atop, and two below. The optional front foglights were now rectangular.

The shaker was gone; in its place, an optional $20.25 sport hood. This had two forward-facing shallow scoops. Many had come to regard the old shaker as adding unnecessary weight and cost. Half of 1972 Barracuda buyers selected 'Cuda, with its standard 340. This implied overall motor selection of 4.45 per cent with 225 I6 power, 42.5 per cent on the 318 V8 and 53.05 per cent favoring 340 V8. In 1967 pony cars hit peak market concentration at 10.5 per cent. A low point in early 1972 of 4.2 per cent was experienced, with slight recovery to 4.3 per cent; of that 4.3 per cent, Plymouth claimed Barracuda had 4.3 per cent!

All the above suggested a smaller market for pony cars, with remaining buyers seeking a plusher, more powerful (or, at least, what power remained), handler. Rather than build sheet ordering, buyers now wanted set menu selection, and plain Jane strippers were out. This explains the rise of 'Cuda, Boss 302, Z/28, Javelin AMX and, especially, Trans Am; often with more luxo content and exterior decoration, but with tamer engines – not the street racers of yore, but possessing base powerplants with less cost for gas and insurance. In June 1970, *Road Test* stated "... mention of the 426 Hemi, or the 440 on an application is like wearing red long johns in a barnyard."

'Cuda, and others, were indeed becoming suburban status symbols, now just as important at the golf club as the drag strip; popular with an increasingly wide range of buyers. However, the official word according to Detroit's demographers was that by the end of '72 MY, the core buyer of ponies was still a man or woman in their early 20s. In any case, Plymouth had profited at this hour from the misfortunes of GM and Ford. There was the April 1972 Norwood, Ohio F-body factory strike, drastically limiting Camaro and Firebird availability.

Henry fell foul with the feds, concerning emissions tuning of his '73s. Thus, Chrysler Corp had an assist with E-body. Barracuda steadied on 18,450 in 1972, and rode the pony car upturn to 22,213 in 1973 sales. Recognising buying patterns, Barracuda's 1973 engine choices were limited to the 318 and 340 V8s. That year, front disk brakes became standard for the first time. Barracuda came with manual disks, 'Cuda with power assisted disks. In 1973 MY the V8s received induction hardened valve seats, so no-lead held no fears.

For driveability and emissions, all Chrysler Corp '73 MY engines had electronic transistorized ignition. Barracuda's new electric choke and heater element reduced choke action when temperature was over 68F. A leaner mix meant less pollution. 'Cuda was still the performance model, so it alone could have a special locked 3.55 rear axled differential, 26in high radiator and fan shroud pack, along with an oil cooler for the optional power steering's pump.

One could have comfort tilt-adjustable steering, but no fast glass or power door locks. For '73 MY,

In charge of the trusty Hurst Pistol Shifter controlled Chrysler A833 four-speed, one could pretend to be Ronnie 'Mr Four-speed' Sox or Dan Gurney. This 'Cuda 340 is a numbers matching car. (Courtesy Ivan Goldenberg, youtu.be/tk3YgEGX3yY)

new-design windshield wiper blades, combated smearing and streaking. They were also better on windlift. Exterior thermosetting acrylic dispersion enamel paint was more durable. '73 Barracuda also had better seat comfort, with a molded-in-place urethane foam base and seatback. A lever-type seatback release replaced the previous push button, and, for trunk security, a stronger decklid lock lived in a steel housing.

Less welcome were the arrival of federally-mandated front 5mph bumpers; these extended bars had large rubber bumperettes. Their extra weight required bracket reinforcement, and plastic color-keyed beauty concealing strips; not bad looking compared to imports, but they increased length to 193in. 1973 width and height rested at 75.6in and 50.9in respectively. Weight for a 150bhp 318 basemobile was a still reasonable 3230lb. 1973 was a bumper industry sales year. Barracuda picked up to 22,213 units.

It was still a far cry from the psychedelic, illustrated, Hemi 'Cuda

Well, if the Chevy boys could say "Camaro by Chevrolet," then why not 'Cuda by Plymouth? In the pony car wars, 'Cuda duked it out with Z/28 and Boss Mustang mightily! (Courtesy Ivan Goldenberg, youtu.be/tk3YgEGX3yY)

'Rapid Transit System' "If you can't beat the system – join it!" ad of 1970. Now, Barracuda was nicely mellow. Good '73 sales provided funds for worthwhile '74 MY upgrades. First, the not-so-good news: the mighty fish got both front and rear 5mph bumpers – taking length to 195.6in – and one year only safety Interlock ignition manipulator made one buckle up, or else ...

Barracuda base price was now $2947. Publicity material promised a 170bhp 2bbl 318 V8. This was 'Cuda's new base motor. A four-speed could be special ordered with said 318. Some 'Cuda 318s got the four-speed via a factory snafu. In keeping with industry trends, radials arrived! With suspension softened to suit optional polyester, in the '70s, you wore it and rode on it; cord steel-belted radials came in FR78-14s. With a guaranteed 40,000 mile warranty, tread life was a bigger selling point than radial's handling/ride benefits.

1974 changes brought longer lasting disk brake linings, and greater claimed engine power via air cleaner fresh air snorkels. Having no external carb bowl vents was claimed to improve cold weather starting. Improved driveability

In an earlier incarnation, this 1971 'Cuda was a bracket racer. A fitting background, given Mopar drag racing heritage. (Courtesy Greg Foy)

Formerly capable of running mid- to high-tens, the present owner is converting this '71 Cuda 440 to street racer form. Part of the roll cage, and monster rear meats will be retained. (Courtesy Greg Foy)

in general was claimed to come from solid fuel type carbs: these had a leaner mixture for emissions. Related to helping '74 Barracuda run better and pass smog law was OSAC, or Orifice Spark Advance Control Valve. This involved vacuum lines between carb and distributor.

For enthusiasts, 'Cuda had a new top engine package. The four-barrel 360 was reconfigured into a 245 horse high output application. This motor replaced the 340 V8. The new package involved a baffled oil pan, shot peened crankshaft with HD bearings, chrome-plated exhaust valve stems, dual valve springs with surge dampers, big camshaft, special valve keepers, rocker arms, bearing clearances and slip drive fan. This hi po 360 even had a dual exhaust system; so did the '74 Mustang II Mach I V6, but that was their only similarity! The 360 in its various forms would be Chrysler's performance motor until the reborn 21st century Hemi V8.

Beyond this, there was even work on a new Barracuda. The 1975 Plymouth Barracuda concept car was based on the E-body, and was created by the Plymouth Advanced Studio. Partly Superbird-inspired, the car was a blend of two proposals. The first from Shunsuke Matsura (a.k.a. Matty) and Don Hood, was muscular. John Sampson and John Herlitz's suggestion was more flowing.

The teams worked separately. They then studied and discussed the proposals to create a blend of the two. Four clay models were made. The final car used the '69 'Cuda 340's alloy rims, and there was a strong Pontiac Trans Am resemblance from the rear ¾ view. There was even a front fender heat escape outlet. Sadly, the depressed state of the pony car market, this time from the recessionary effects of the '73 Gas Crunch, meant there was no '75 Barracuda. 1974 sales had plummeted to 11,734. With the Big 4 scaling back, or leaving pony cars altogether, Chrysler Corp had little choice but to follow suit.

Hope for the future has arrived in the form of a V6 turbo Alfa Giulia-based Barracuda prototype. This coupe was shown to dealers on August 25th, 2015. The demise of the Plymouth brand in intervening times has made re-branding challenging.

Dodge Challenger 1970-74: the fish's friend

With the arrival of the E-body Barracuda came a Dodge sibling called Challenger. Like the relationship between Valiant and Dart, the Challenger lived on a larger 110in wheelbase. 1970 dimensions were 191.3in length, 76.1in width and 50.9in height. The Challenger's design came from the pen of Carl Cameron, who also designed Charger. Indeed, Challenger's four circular headlamp grille was originally designed for the proposed gas turbine-powered Charger. The Challenger design was finalized in 1968. The coupe was subsequently built at Chrysler's Hamtramck, Michigan and LA Assembly (Maywood), California facilities.

Versus Barracuda, the Dodge

This 1970 Dodge Challenger was specified with the 1970-only A66 package. Factory fitted were 340 V8, performance hood tie down pins, HD four wheel drums, and HD suspension. (Courtesy Brian Horvath)

It's hard to not want to make a good thing better; in this case, substituting a 440-6 motor with aluminum heads. The 4.10 rear axled Positraction lsd makes this Challenger a formidable drag strip opponent. (Courtesy Brian Horvath)

Factory optioned with TorqueFlite auto, full instrumentation Rallye Gauges, and interior light package. The black interior matched the black vinyl top. (Courtesy Brian Horvath)

Challenger cost 88 bucks extra. It was 4.6in longer and weighed 125lb more, but had some welcome extra rear legroom. A four circular-headlamp face, and single tailwidth lens cover distinguished Challenger. There was also a mid-fascia back-up light. Barracuda's front foglights couldn't be adapted to Challenger. The two siblings shared no sheetmetal, except for the roof. This was criticized later, but at the time, management felt the appearance-conscious nature of the pony car market and potential high sales volume, warranted the extravagance.

With the interior and engine line ups, Barracuda and Challenger were close. It was the same dash molding, and even though Challenger was in the same upscale Cougar, Firebird crowd, the same press and public qualms. *Motor Trend*'s John Fuchs mentioned a stark interior with crassly finished plastic, and like Barracuda the doors kind of clanged, rather than thunked.

Challenger SE, R/T & T/A

Hardtops and ragtops were on hand, with three distinct trim levels. There was base Challenger, the upscale Challenger SE and the high-performance Challenger R/T. The 198in^3 I6 was omitted for larger Challenger. Business started with the 225 Slant-6, reaching 330 horse 4bbl 383 V8 for regular Challenger. Challenger SE was equivalent to Barracuda's Gran Coupe. That meant six-way manually adjusted driver's seat, leather interior, overhead consolette with warning lights for door ajar, low fuel and seatbelt use, vinyl roof and smaller, formal side windows.

The performance variant was the equivalent of Plymouth's 'Cuda. As with other hot Dodges, R/T stood for road and track. Power commenced with the 335bhp 4bbl 383, or NCO 340. Then came the 375 horse 440-4, 390bhp 440 Six Pack and the 425 pony Hemi 426. Just like 'Cuda business. Challenger R/T had a 150mph speedo and eight grand tach, belonging to the Rallye instrument package. In 1970 it was possible to order base Challenger and R/T, with SE's luxo

This Challenger was ordered with white Bumble Bee stripe, but the decorative wheel lip moldings were deleted. Factory power front disks have been retrofitted. (Courtesy Brian Horvath)

trim level – and order the cars they did. The public purchased 76,935 Challengers in 1970; the pricier Challenger outsold Barracuda three to two. Of that total, 3753 were R/T hardtops in SE trim, and 963 R/T ragtops rolled off the assembly line.

Challenger T/A was visually distinguished by a matte black fiberglass hood with forward-facing hoodscoop. A pinstriped, solid-black 'T/A' decal ran above the body's profile character line, from the front fender leading edge, to the first third of the rear fender, between a fiberglass front spoiler and fiberglass ducktail rear spoiler. Asymmetric tires were sized like AAR 'Cuda 340. Also like that 'Cuda, there were hood pins and side exhaust system; that is, low restriction duals reaching stock muffler, then to the side, as a chrome tipped megaphone outlet. Nothing like a jacked-up rear and side trumpet to announce street presence!

It was all good enough to go racing. So it was that 'Cuda had an SCCA T/A cousin, in the identically powered 1970 Challenger T/A 340. The T/A stood for the SCCA's Trans Am series. The motor was the 340, with triple deuce carburetion, and grossly underrated 290 horses. All to homologate Sam Posey's racer. The car in question was the No 77 'Sub-Lime' colored Challenger T/A 340. Dodge contracted Ray Caldwell's Autodynamics of Marblehead, Massachusetts, to run the team. Keith Black did the car's V8.

Dodge's effort yielded a couple of top-threes. The best overall Mopar result came at Elkhart Lake; Sam Posey led the race, but a slow pitstop handed victory to Mark Donohue's Penske Javelin. Swede Savage came second with Posey third. Dodge managed 18 points in the year's championship. As per Plymouth, oil starvation limited engine durability in long races. There wasn't the time nor budget to develop the downsized 340 optimally for this race format.

A major disagreement eventuated between Caldwell and Posey mid-season; this caused Carroll Smith to take over running the Dodge team. Posey's No 77 racer was built from a showroom floor Challenger T/A 340. A second No 76

This is a numbers matching '71 Challenger 383 in EL5 Butterscotch Yellow: a shade unique to '71 E-bodies. (Courtesy Matthew Michalec, mattmichalec.com)

car was constructed mid-season, from a Dan Gurney AAR team provided chassis. Sam Posey then alternated between No 77 and No 76, except for the final two races. Here, Ronnie Bucknum drove No 76 at Seattle, Washington. Tony Adamowicz took over for the Riverside, California round.

Dodge was out of SCCA T/A racing at the close of 1970, but did plan a 1971 road-going Challenger T/A 340. Ads were done, but the model wasn't released. Similarity, once again, with Plymouth, in that you could build one from the option sheet. 1971 was a year of some omissions and reductions. A ragtop R/T was off the menu. The 2bbl 383 V8 was down from 290 to 275bhp. Revisions to motor choices mirrored that of Barracuda and 'Cuda. So did the sales slide. 1971's total fell to 26,299. Challenger always had a sales edge on Barracuda, but E-body was in trouble.

1972 saw expected range rationalization. No big blocks, leather seats, fast glass or convertible option. However, one could now get a factory sunroof. It was in keeping with the A-body's 'Convertriple' phenomenon. Stemming from the fear of federal rollover testing, with all the regs raining down, many enthusiasts wanted to roll over those pesky feds! The Challenger R/T was replaced by the Challenger Rallye. It was the new performance edition. No shaker, nor Rallye id badges, but a faux fender brake vent was at hand, plus that trusty 340 V8.

Challenger's cosmetic changes were always accepted. In '71 MY a split front grille, and separate left and right side backup reversing lamps arrived. For

Given the Challenger R/T ragtop's 1971 absence, this 300 horse 4bbl 383 car was top convertible V8. This was an H51 a/c car; the system will be re-installed during restoration. All this Magnum power goes through a 3.23 rear axle ratio Sure-Grip lsd. (Courtesy Matthew Michalec, mattmichalec.com)

This 1971 Challenger 383 convertible was built at the Hamtramck factory on January 26th 1971. Challenger's 2in longer wheelbase provided a more commodious back seat than Barracuda. (Courtesy Matthew Michalec, mattmichalec.com)

Actor Bill Bixby drove a '70 Challenger ragtop in Aaron Spelling's 1971 TV movie *Congratulations. It's A Boy!* 1971 was the final curtain for E-body convertibles. (Courtesy Matthew Michalec, mattmichalec.com)

1972, the extended, sub-bumper 'Sad Mouth' grille joined a revised tail with four individual taillight pods. 1973 and 1974 changes, concerned the adoption of federally mandated impact bumpers. 1972's flush bumpers, with small optional front guards, gave way to an extended 1973 front bumper, with standard large guards. The final-year '74s just added large rear guards to please the feds.

On the subject of distinguishing Highline Challengers, West Coast Dodge dealers tried to do just that, with some small block 'Western Sport Specials.' Pedal dress-up kits, vinyl roof, rear valence exit exhaust system, rear-quarter panel "Western Sport Special" applique and a few with interior vacuum operated trunk release. Dealers trying to steer prospects to their corral. Being California and San Francisco dealer available, it was natural they come from the LA factory, but some WSS 340 cars from Hamtramck exist.

There was also the late Hamtramck and Van Nuys production stripped 'Deputy.' Based on the entry level Challenger Coupe. With fixed rear-quarter panes perhaps, but this Deputy was no dog. It was an arresting way to get a 383 V8 three-speed combo. With that cop favorite V8, one could pretend to be the long arm of the law!

E-body in show business

Barracuda and Challenger played some great roles on the small and silver screens. For continuity, some late '71 Barracuda ragtops were given '72 MY front and rear fascias. This was so the rides would look current on Paramount Studios TV shows *Mannix* and The *Brady Bunch*. The same was done for Challenger convertible, concerning TV's *The Mod Squad*. The car's fender tag codeline with an 'R' could mean a TV promotions car. Most famously a white 1970 Challenger R/T 440 Magnum co-starred with Barry Newman in the 1971 car chase spectacular movie, *Vanishing Point*.

Less well known was the use of a '70 Challenger convertible in the TV movie. '*Congratulations. It's A Boy!*' In this 1971 Aaron Spelling production, where Chrysler supplied the vehicles, the main character Johnny Gaines drove said ragtop. Gaines worked as a sales exec in his dad's womens' dress company. Bill Bixby played the carefree middle-aged batchelor, who suddenly encounters the 18-year-old son he never knew he had, on his doorstep. The chickens had come home to roost, and the hedonism of the '60s was over. The pony car market experienced a very similar thing.

In the 1990s, Don Johnson's TV police inspector, *Nash Bridges*, drove a '71 Hemi 'Cuda convertible. Several cars were built for the TV show, but not one had a Hemi V8. In the movie *Terminal Velocity*, Charlie Sheen's skydiving instructor drove a white Challenger hardtop. Unfortunately, it was shot up to destruction. The use of a white Challenger was a tribute to *Vanishing Point*. *Terminal Velocity's* director Deran Sarafian was the son of *Vanishing Point* director Richard C Sarafian.

In terms of a classic show with a classic ride, TV detective *Mannix* drove a '74 Challenger Rallye during the 1973-74 season. This was a special order car, made for the show. It carried all options, including 360 4bbl performance motor and the rare factory sunroof.

Any '74 Challenger is a rare beast, only 11,354 were sold that year. This followed 1972 sales of 22,919 and the 1973 upturn to 27,930 units. In November 1972 *Motor Trend's* Jim Brokaw acidly asserted that commercially, Challenger "... challenged hardly

This is a 1972 Dodge Challenger Rallye with a 4bbl 360 V8. The sporty Rallye package was a replacement for 1970-71's Challenger R/T. (Courtesy Edward Hessel)

anybody." Most cars were sold in the honeymoon debut year. A total of 165,437 Challengers were made, and production stopped in mid-1974.

Challenger showed the commercial trends of the time. Like Barracuda, sales were cannibalized by cheaper A-body Duster and Demon/Sport below, and from Charger above. Chrysler's rivals experienced similar in-house sales poaching. In 1970-71 pony cars lost sales to normal compacts, sub compacts and economy cars, as American buyers seeking smaller vehicles, found cars closer to their needs.

1972-73 saw pony cars rise as midsize and full-size buyers, downsized to fight inflation, and just to get a more manageable size car. Ponies were the new happy medium. They went from street racers, to a sports/luxo blend in a smaller size, that buyers have sought ever since. In spite of falling sales, Challenger R/T held steady on 3814 units in 1971. At 8123 sales, Challenger Rallye made up over one third of 1972 Challenger sales. Now, sport and lux were joined at the hip. This partly explains the difficulties Cadillac, Imperial and Lincoln have had staying relevant in the years since.

Motor Trend's John Fuchs described the pony's progress in the April 1974 Fearsome Foursome test: "... that blanket of uncertainty generated by the Energy Circus will very likely serve to resurrect the entire pony car population – not as zoomy, pseudo-race-cars, but as small sporty sedans that are quite economical once you choke down the purchase price." With 0-60mph in 9.6 seconds and 15.5mpg overall, the Challenger 318 on test was efficient. It was quicker than the 400in^3 Firebird Esprit, and less than 0.2 seconds slower in passing tests than Camaro Z/28. Challenger's ride/handling compromise was also judged good.

Fuchs remarked that Chrysler's mechanical components and engineering were superb. Challenger, Camaro, Firebird and Javelin, on test, were judged to have that special sporty car feel. So when buyers rediscovered ponies in 1973, they were certainly wanting a smaller, efficient, and maneuverable package. John Fuchs also mentioned *Motor Trend* had wanted to test a 'Cuda, but was unable to procure one.

The observation was also made that Challenger was the oldest car on test, and in need of a redesign. Even as late as the mid '70s, American buyers still looked for the latest thing. By the

The Challenger was built at Chrysler's Hamtramck, Michigan and LA Assembly, California factories. (Courtesy Edward Hessel)

standards of the day, the four-year-old E-body was getting long in the tooth. It seemed fickle fashion saw buyers flock to midsize personal cars, like Monte Carlo, Gran Torino and the new Matador Coupe. All did good business in 1973-74, while even GM F-body floundered. All in spite of the gas crunch.

E-body fared little better in Europe. Chrysler exported CKD kits to Swiss firm AMAG Automobil und Motoren AG, located in Schinznach-Bad, near Zurich. The company carried out final assembly, and made sure Challenger met Switzerland's strict pollution and general vehicle laws. It was an exclusive operation with few made. Throughout Europe, GM F-body and Mustang, stymied Chrysler E-body's commercial success. Henry's pony had done brisk business in France. Dodge hoped Challenger could emulate this.

The Challenger was exported to France, and sold through Chrysler Simca France. Sadly, Challenger couldn't make inroads, given Mustang's already established fame. Racing was another matter. Cousin 'Cuda was successfully raced by Chrysler Simca France's factory backed team. Its Hemi 'Cuda was driven by team director Henri Chemin, between 1970 and 1973. Four french Group 1 class championships were won. Three on the track, and one for hill climbing. Monsieur Chemin then sold said Hemi 'Cuda, to friend Jochen Mass.

It wasn't easy selling E-body at home either, especially the ultra hot versions. The Big 4's racer facsimilies always moved off the showroom floor slower than molasses in an ice age. In racing, the Hemi was respected; in the showroom it was hard to sell. $1228 extra to get a Hemi Challenger in 1971 was a lot of bread, but some got freebies. A 1971 Dodge Hemi Challenger was the last car received by legendary speed-record setter, Norm Thatcher, from Chrysler Corp.

Racers knew Mopar hardware could be trusted. There was plenty of Chrysler connection to the real life Cannonball Run road race. Started by *Car and Driver* journalist Brock Yates in 1971, it was a social protest against blanket speed limits; proof that speed alone doesn't make one unsafe. Brock Yates did the original race in a Dodge van. Starting from Red Ball Garage in Manhattan, the van took 40 hours and 51 minutes to cover 2950 miles, and crossed the finish line in Redondo Beach, California. A 15 year old Brock Yates Jr helped with route navigation.

Dodge's Challenger enjoyed the commercial distinction of outselling its Plymouth sibling three to two. As per Barracuda, there were no big blocks post 1971. (Courtesy Edward Hessel)

In late 1971, Brock Yates received a challenge from the PRDA (Polish Racing Drivers of America). Some irony, here, in that the central character in the movie *Vanishing Point* was Kowalski, the Challenger R/T delivery driver. In the second Cannonball Run, Oscar Koveleski's hopped up Chevy V8 van was defeated by Yates and Dan Gurney's Ferrari Daytona. Gurney was no stranger to the 'Cuda and SCCA Trans Am racing. Fewer know that Brock Yates competed in the same series earlier, driving a Dodge Dart!

In 1999, 20 years after the final Cannonball Run, came the Cannonball Classic: a slightly more sedate revival of the previous concept. Organiser Martin Swig brought a 1951 Cunningham C4R road racer replica. This Le Mans machine was powered by a Hemi 331 V8. Two Dodge Challengers also took part, one of which was Brock Yates' own 1971 hardtop, which took part in the 1972 and 1975 Cannonball Runs. In the 1999 Classic, it was driven by Brock Yates Jr. Apart from aftermarket instruments, this '71 Challenger had a radar detector, mobile phone, aftermarket buckets, 45 gallon fuel cell, stiffer suspension, and quick ratio steering.(5)

The other Challenger was a 400 horse machine. Alongside a Datsun 240Z, it was part of a two-car, Tokyo-based team, which included popular Japanese car magazine *Options* founder Daijiro Inada. The 1999 Cannonball Classic was won by Wes McNay and Franz Lorist, in an orange Plymouth Superbird. The 1970s were a commercially challenging time for Chrysler Corp. The market may have been a tough customer, but whenever Mopar challenged the road, it won!

Dodge supplied eight 1970 Challenger R/T 440 Magnums for production of the 1971 movie *Vanishing Point*. By the close of filming, only this car had survived! (Courtesy Floyd Garrett, musclecarmuseum.com)

3

Dart – From Senior Compact to Muscle Car

Starting with standard-size

Chrysler bought Dodge from the Dodge brothers in 1928. The marque was placed upscale of DeSoto, and beneath Chrysler. However, come 1933, DeSoto was elevated to premium status, under Chrysler. A further reversal came with DeSoto's de-emphasis and demise. This led to Dodge going upscale once again, with more lavish Plymouths taking up their vacated positions. In the wake of the late '50s recession, there was much corporate restructuring, plus a new junior brand called Dart.

From the 1930s, Dodge dealers had sold plain Plymouths; so, too, selected DeSoto and Chrysler-Imperial outlets. The future would show the supermarket 'mile-of-cars' dealership would win out. In the meantime, Dodge went exclusive. Instead of being a medium-price brand which sometimes sold low-price cars, it now had a popular-price model of its own. Dart was a shorter standard-size family car, built on a 118in wheelbase. It was a part-unibody design, with a subframe for engine, suspension, and sheet metal forward of the firewall. Pure unibody lay rearwards.

This made for easy repairs and towing, along with more interior space. The latter came from a lower floor, that avoided separate chassis frame rails. The new Dart involved minimal disruption and cost to Chrysler Corp's existing production set-up. As a name, Dart was a 1957 Ghia show car, and product planners wanted this moniker for the 1960 low-cost line. Management wanted a detailed name research study, which literally turned up 'Zipp' – that was the name the study offered! Wisely, the original choice of Dart was reinstated; a rose by that other name wouldn't have sounded too sweet.

Dart hits the bullseye!

Dart turned out to be Dodge's savior. 1960 sales hit a record 367,804, and pushed Dodge to sixth in the sales race. All this in the immediate aftermath of the late '50s recession. A low, or 'popular' price (as Dodge put it), standard-size car seemed to be what Joe Q Public wanted; not the high-priced, fancy Edsel; not even the normal, full-size Dodge Matador, and upscale medium-price cousin called Polara. These two models rode on a 122in wheelbase. So did Dart wagon, however Matador and Polara only accounted for around 41,000 Dodge sales. It seemed that Dart poached prospects from even Plymouth showrooms.

As Dodge moved from 11th to sixth, sales in 1960 were up 107 per cent. 85 per cent of 1960 Dodge sales were down to Dart. It was even a nice car, with technical touches absent on Big 4 rivals. Dart's three trim levels, going upscale, were Seneca, Pioneer, and Phoenix – Seneca put one in the Chevy Biscayne, Ford Custom, and Plymouth Savoy ballpark; Pioneer raised the stakes to Chevy Bel Air, Ford Fairlane, and Plymouth Belvedere; Phoenix entered the rarefied tow zone of Impalas, Galaxys, and Plymouth Fury.

Slant-6, torsion bars & ram induction

One couldn't have a six-cylinder Dart convertible, but an I6 Phoenix was possible. That said, it was hard to call a Dodge dealer a low-price outlet due to Dart. Cars in the family standard-size segment were getting jazzy. With all the options, one could make an Imperial owner do a double take! Aside from sedan, hardtop, and ragtop, wagons could roll with six or nine passengers. These Windsor, Ontario-built Darts were rather roadable. With the front torsion bars and rear leaf-sprung axle (Chrysler style, since 1957) Dart offered a good ride, and rough-road handling akin to a sports car.

With Chrysler's all-around Oriflow tube shocks, a 1960 Dart got airborne when *Hot Rod*'s staff were trying said machine at Chrysler's proving grounds. The car made a controlled landing.[6]

Plus, with Dart sitting lower, and being narrower and shorter than Plymouth's equivalent, handling had come to the fore; even more so, with the new 225in^3 Slant-6 – a newly launched version with Chrysler's six converted to an overhead valve layout, advantages abounded. Inclined 30 degrees, it gave Dart a low hood line, and 52/48 per cent front to rear weight distribution.

There was nothing low-cost about Dart performance. Its first optionable motor was Seneca/Pioneer's 1955-introduced 318 V8. It made 230 horse in 2bbl trim, and 255bhp, at the same 4400rpm, in hi po four-barrel form. However, the Big Daddy 1960 MY Dart option was the Phoenix-only D-500, or Super Red-Ram 361 Ram Induction (RI) V8. It cost $413 and delivered mid-16s for the 4100lb plus Phoenix. Power was 310bhp at 4800rpm, with 435lb/ft at 2800rpm. The 10:1 CR requested premium gas, and the AMA (American Manufacturer's Association) demanded discretion. Dart publicity didn't state the power benefits of RI. There was an anti-racing Big 4 industry pact afoot, but folks knew Mopar had it going on.

In March 1960, *Car and Driver*'s man wore a humorous ram outfit, and placed his hooves on the hood. *C&D* said RI would make other performance hounds sheepish when a super Red-Ram Dart arrived at a street Christmas tree. D-500 Dart was top dog at the traffic light drags. Ram Induction was an intricate, long-runner intake manifold system. A forerunner of late model fuel-injection set ups, except using super long tubes and carbs. Two Carter 4bbl AFBs placed 4ft apart, with a throttle linkage in the center of a criss-cross cast aluminum intake manifold.

Manifold diameter was 1⅞in by 1⅛in, and there lay around 30in between carb venturi and the intake valve. In effect there were eight separate manifolds. The system used the racing know-how pressure wave effect. It needed such long runners to get usable performance gains at street rpm. A partial vacuum effect arose when the intake valve opened. Such gentle supercharging helped fill the combustion chamber 10 per cent. An indication of the effect was shown by the Dodge Polara's two 383s. The regular 4bbl 383 made 325 horse at 4600rpm and 425lb/ft at 2800rpm. The dual 4bbl D-500 RI 383 made 330bhp at 4800rpm and 460lb/ft at 2800rpm. So it was all about the torque.

No special gas beyond premium, nor driver intervention needed to get the gain. However, underhood engine access became a real bear. The sparkplugs did reside under the exhaust manifolds. According to *C&D,* one could special order the D-500 383 on Dart Phoenix, but general belief was that 383 was Polara-reserved in 1960, regardless of RI. Plus, one could have a three-speed stick with D-500 Phoenix. According to Chrysler division engineering chief Burt Bouwkamp, the late '50s three-speed column manual, with synchro on second and third, was intended for high-speed, full-throttle Bonneville record runs in top gear.

First and second were very narrow gears, that would get burnt up with full gas given. Bouwkamp said they screened the few manual transmission buyers, to make sure they knew the score. All 118in and 122in wheelbase D-500 cars, had a 3.31 rear axle ratio. D-500 RI machines also had a special HD TorqueFlite, with increased oil pressure to reduce clutch wear and slippage during pedal-to-the-metal upshifts. 1960 was a transitional time from Dodge's Mobil Economy Run feats, and general thrift, to something livelier.

Car and Driver's March 1960 issue mentioned the "sports-dragster," "... one which wins trophies on Sunday and takes the kids to school on Monday." It seemed Mori Q Public had just decided he wanted to become a racing driver! In spite of the AMA's ban on fun, there were signs money could be made from performance options. In November 1959, Ray Brock stated in *Hot Rod* that the auto industry was the biggest sector in America. Dart certainly helped Dodge and Chrysler Corp join that crowd.

Dodge in 1961

1961 held promise, and there were promising Dart developments. Chief upgrade was the availability of 383 RI motor in the Dart range, no longer a Polara exclusive. There was a new HD manual transmission, for 361/383 cube V8s. Plus, the refinement of a new power steering pump mounting, and extra suspension leaf on I6 cars. The only pickle in the ointment was Virgil Exner's Dart restyle. The legendary Exner, who had introduced Chrysler Corp to tailfins in 1955, had gone too far in the eyes of many. The front concave fascia was not liked, but it was the new tailfin treatment that rankled most. The idea was to reduce rear-end heaviness. However, a round fin with the highest point forward, rather than rearwards, was a touch avant garde, and worthy of Salvador Dali.

Functionally, the small, rear, low-mount taillights were difficult to see. Mid model year, auxiliary round taillights, inboard of the fin, became a dealer option. At this time the aluminum die cast block 225, also became available. Plus, all '61 MY Chrysler Corp cars had an alternator, which was standard on the 1960 Valiant. As a driving machine, Dart was as good as ever. Terrific power teams, generally solid engineering hardware and great balanced handling. Commercially, however, Dodge hit the skids in 1961.

Dodge was down to 269,367 units, and ninth in the sales race. Valiant's essentially badge-engineered cousin, called Lancer, sold just 74,773 units, versus 143,078 for the Plymouth original. Of the 142,708 1961 Darts sold, almost one half were Senecas. Plus, Matador had said its final 'ole!' at the end of 1960. So it seemed the public were still appreciating, the concept of an affordable standard size car. Chrysler Corp then took another gamble, feeling rivals would go downsize full-size for 1962.

In 1961, even the smog was happy! However, public and critical opinion weren't too warm concerning the '61 Dart restyle. Specifically, Virgil Exner's latest tailfins. (Marc Cranswick)

Dart goes midsize

The all new B-Platform midsize Dart, certainly lived up to its "New Lean Breed" slogan. Visually, it was mistaken for Lancer. It's long-hood, short-deck styling was considered exciting, sporty, and European. The 1962 Dart rode on a 116in wheelbase. At 202in long, it was 7.4in shorter than 1961's Dart. At 76.5in wide and 54.1in high, it was 2.2in narrower and 0.7in shorter as a sedan. Weight saved was up to 200lb per version. The emphasis was on efficiency. Like a new, lighter TorqueFlite auto for V8s. With a one-piece, die-cast aluminum housing, its torque converter had greater capacity and was attached to a flexible drive plate, not the crankshaft.

The new Dart was a full unibody. Its lithe nature, more rigid structure, and torsion bar/leaf-sprung live-axled chassis, promised and delivered great handling. Although Dart would enjoy this platform for just one year, the B-platform would prove a winner. It finished with the 1981 R-bodies. '62 Dart came as a two or four-door hardtop, four-door sedan, wagon, and convertible. Trim levels were Dart, Dart 330, and upscale 440 with rear-fender hash marks; not to forget extra-fancy Polara 500. A 305bhp, 4bbl, 9:1 CR 361 V8 was the top motor. For performance on a budget, the hi po 318 came with duals, a four-barrel carb and 260 ponies. The 1961 383 D-500 RI motor had made Dart a 16 second flat car. It was speculated the new model's compact, low hood arrangement made ram induction hard to fit.

In 1962, the Dart became a very roadable B-platform midsize car. Available in 23 flavors on launch, the most exciting was this mid '62 MY Max Wedge 413in^3 engine package! (Courtesy Floyd Garrett, musclecarmuseum.com)

Dodge re-engineered the Ram Induction system into a more compact form, to fit under the low hoodline of the midsize '62 Dart. Sitting on the Max Wedge 413 V8, it produced a 410 horse holy grail! (Courtesy Floyd Garrett, musclecarmuseum.com)

The success of Valiant led to this 1961 Dodge Lancer Sport Coupe sibling. Note the rear window Rat Fink decal. (Courtesy Larry Williams)

A nice car, but, sadly, Lancer sales were only a little over one half of Valiant's tally. Dodge had to differentiate its car better. (Courtesy Larry Williams)

Ramcharger 413!

Chrysler engineers were indeed designing a more compact RI set up for midsize cars. It would all be part of a Plymouth and Dodge available 'Ramcharger 413' $374.40 package, which turned one's family sled into a racer. A complete factory-made engine/suspension package, riding on Goodyear 7.50-14s, that delivered sub 6 second 0-60mph and mid-14s at over 100mph! On top of the Max Wedge 413 RB big block, lay the revised RI apparatus. Still two Carter AFBs at a diagonal distance, but with runner length reduced to 15in.

The 413 featured a special free flow head casting, and oversize 1.88in exhaust valves; 1.60in had been normal. Larger ports, HD forged 11:1 CR pistons, and a modified version of the 1961 RC-92 camshaft. The big cam now had 300-degree duration, with over 0.5in of lift. All the conventional racing tricks of the era: bigger piston and bearing clearances, hardened crank, HD tri-metal bearings, magna fluxed conrods and baffled oil pan. The dual point distributor had centrifugal advance alone. This was a no-compromise light solid lifter V8 with HD oval springs/keepers, 6500rpm redline and 1000rpm idle speed.

This Dart never got exhausted, not with 3in headpipes, capped openings, reverse flow mufflers, and cross-over pipe as a balance tube. As an automatic, a Dart Ramcharger 413 weighed under 3500lb. To reduce weight, the steering box and automatic's cover were made of die cast aluminum. The standard package had no heater; given the AMA's anti-racing pact, this was probably to discourage kids using the heater to make the engine run cooler and win street races! The HD TorqueFlite helped in racing. It had zero slip, and was engineered to upshift at 5600rpm on full throttle.

Use pushbuttons one and two, and Ramcharger 413 would stay in those gears. Uncap the exhaust to get a genuine 70 extra horses, and a lot of noise. One could fit racing tires, and go for the optional 13.5:1 CR 413. The result would be low-13s at over 110mph! Chrysler's big V8 manual three-speed had been upgraded for 1961. However, shifts still weren't fast or positive enough for the strip, so, during 1962, there was a production change to a Warner T-85 HD three-speed with Hurst linkage. Standard fitment was a Sure-Grip lsd, with 3.91 rear axle.

410 horse at 5400rpm sounded amazing. It went through an HD driveshaft, with a six leaf sprung live axle as part of the HD suspension. Unfortunately, the price of 21st century performance was a cantankerous street nature. Ramcharger 413s behaved like the 1968 Hemi Darts and Barracudas. The cold-natured aluminum intake manifold made for rough cold weather idle. Indeed, the manifold was designed to perform over four grand, exacerbated by big valve overlap. A lack of low rpm spark advance, and loose piston clearances, caused high gasoline and oil consumption, respectively.

The aforementioned high idle speed, made for brake creep in traffic. Indeed, scratch 'great' and substitute 'scary.' This little lady was going to take off by itself, just like that demonic Fury Christine! Or, in the words of *Motor Trend*'s Roger Huntington: "This is a 'racing' car – not a street car. But what a racing car ... and what a car for the money!" Understandably, a whole new generation of drag racers got turned on to Mopar muscle thanks to Max Wedge and Hemi V8s. This was the case with Dandy Dick Landy.

Ramcharger racers

In the early '60s, Dick Landy made his way to Mopar, attracted by the Max Wedge 413. A stock Plymouth 413 outpunched his modified Ford 406, and the rest was Hemi history. In a way, racers were giving racers their rides. The Ramcharger 413 package arose from the work of a group of young Chrysler

Corp engineers called 'The Ramchargers.' Working out of Chrysler Engineering offices in Highland Park, Michigan, the bunch sought solutions to win on the strip. These racing hobby engineers included Tom Hoover, Mike Buckel, Jim Thornton, Dick Maxwell and John Werley.

The Ramchargers got factory access, but not actual support. All hands were on deck working towards the Hemi's 1964 Daytona 500 debut. The team were linked to Hemi's first appearance outside NASCAR, and the start of A/FX, and therefore Funny Cars. The new Ford Thunderbolt 427 was putting the heat on Max Wedge. The new Hemi was 100lb heftier than Max Wedge. It's believed Jim Thornton was behind the move to an altered wheelbase, so that a racing machine could get some traction.

So Ramcharger 'Business Coupes,' with more weight-saving aluminum, and no backseat, had the Max Wedge V8 in an altered wheelbase FX Dodge. '64 MY Max Wedge cars had a full two-door body, with Dodge 330 and Plymouth Savoy moving into spring with Hemi power. Although the cars weren't factory jobs with Hemi VIN, it is strongly felt that the Ramchargers' Dodge was the first drag racing competition car with Hemi power. Before all this the Ramchargers, campaigned a '61 Dodge sponsored by Taylor Motor Sales of Detroit, in the then new SS/S class at the NHRA Nationals in Indianapolis.

The Ramchargers' Jim Thornton, utilizing a Hurst three-speed floorshift, 4.56 gears, and extra-stiff right rear spring, defeated Dyno Don Nicholson's Chevy 409 at Detroit Dragway in April 1962, over two runs. The Ramchargers then teamed up with Elton 'Al the lawman' Eckstrand. Eckstrand was a former corporate lawyer, who did work for the Big 3. He also happened to own a '62 Dart Ramcharger 413 automatic.

'Al the lawman' won the 1963 Winternationals for the Ramchargers, but was a sharp operator, who managed to nab some sponsor money for himself. The Ramchargers split up with Eckstrand, who moved to Plymouths as a free agent. He even pulled strings to get one of the four new altered wheelbase cars.

Norm Thatcher – speed legend

The other place Mopar magnificence could be seen was on the salt flats. Pontiac may have had the Bonneville, but at Bonneville, legendary elder-statesman-of-speed Norm Thatcher was Mopar master of timed runs in all three Dodge class sizes.

At El Mirage Dry Lake, ice cream loving Thatcher took a 1960 standard size Dart to 159.29mph in atmo class. He added a 167.28mph run in the blown supercharged category. Then came Bonneville records, of a two-way flying-run mile average of 164.087 in A Production Gas, with 164.245mph in B Production Gas. Then came the unsuccessful attempt to reach 200mph in the new 1961 Lancer. It was an attempt done at the request of Chrysler Corp to promote the Valiant's virtual twin. Norm Thatcher did reach 183.48mph in Lancer at Bonneville.

More successful was Thatcher's spell in the new midsize '62½ Dart Ramcharger 413, with licence plate XUG077. A blown example was fielded at Bonneville. A two-way flying mile average of 189.826mph in the B Supercharged class.[7] Thatcher collected 12 trophies, and was top racer at Bonneville in 1962. Norm Thatcher even ran a Dodge pickup with the Ram Induction D-500 motor. Such racing exploits were all achieved prior to the AMA lifting its anti-racing ban. However, of greater concern to dealers was the absence of a full-size '62 Dodge.

Dodge goes senior compact

To correct the omission, Dodge cocktailed the 1961 Polara and 1962 Newport, to create the Custom 880. With all this, the compact Lancer wasn't working out. Dodge needed something truly different to Valiant, so the firm came up with a larger, freshly-styled Lancer replacement. It rode on an 111in wheelbase, which made it a king-size compact, or intermediate like the '62 Ford Fairlane.

However, Dart wagons shared the same 106in wheelbase as Valiants. Dart displayed some front-third resemblance to the 1963 Chrysler Gas Turbine car. There was the fashionable tunnel port front fender headlight complex, as seen on subsequent '64 Rambler American and Porsche 911. The curves of the '50s had given way to flat planes and sleek lines.

As a company, Dodge was 49 years old, and offered three model lines: the new Dart was it's smallest offering – then came the popular-price, 119in wheelbase 330, and medium-price Dodge 880. The first two were showroom available by October 2, while Dodge 880 would arrive in December. However, all eyes had nearly been on 'Lancer.' The decision to name the new senior compact 'Dart' was a last minute one; the Lancer nameplate was originally going to be retained. As things stood, '63 Dart came in three trim lines, five bodystyles and nine model versions.

Matters started economically with the 170, available in two and four-door sedan. Deluxe Dart 270 added a convertible, and then came GT. Acknowledging the popularity of sporty imports, Dart GT could be a two-door hardtop or ragtop. Either way, buckets were standard. A triangle GT badge lay on the hood's leading edge, ahead of the power bulge. There was also a GT badge on the front fender behind the wheel well, and 'GT' script on the rear fascia.

For all the fanciness, Dodge General Manager Byron J Nichols advanced the new Dart on solid virtues – that is; dependable, durable, and economical – a reminder that we weren't far on from the late '50s recession. That said, '63 Dart had several optional features announced on debut: power steering, power tailgate window for wagon, a/c, Torque-Flite auto, and power top for ragtop. Above the 101bhp base 170in^3 six, was the optional 225 I6. Dimensions for four-door sedan were 195.9in length, 69.8in width and 54in for height, or, the same as a '60s W108 Mercedes 300SEL. By mid '64 MY, Dart had an optional 180 horse V8 that would match the Mercedes' power.

Going with Chrysler's rediscovered quality, Dart could boast durability improvements and incremental gains over Lancer. Parallel windshield wipers with 1in longer blades, a ventilation system delivering 40 per cent more air, and a new all-transistor radio. Seats were claimed to offer more support and possess plusher, more wear-resistant materials. Dart was dipped in a deeper vat of special primer, body moldings had rust proof clips, the exhaust was

Lancer was too close to Valiant. The new 1963 Dodge Dart senior compact was distinct in engineering and style. The new range was a success. (Courtesy Evan Eatkinson)

1963 Dart included this 170 economy trim level wagon. The optional 225 Slant-6 fitted made 145 gross horses. (Courtesy Evan Eatkinson)

Compared to Lancer, Dodge claimed an improved heat/vent system – faster and more precise heat control, with 40 per cent greater fresh air supply. Dart's all transistor radio, was new for '63 MY. It featured 'Instant Sound' and less current drain. (Courtesy Evan Eatkinson)

For many, the normal full-size 122in wheelbase family wagon was just too big. Dart wagon cut wheelbase to 106in, was cheaper to buy and run, and was still peppy on Slant-6 power. A power tailgate window was optional. (Courtesy Evan Eatkinson)

aluminized, and extra splash shields protected the lower body from corrosion.

Gas tank capacity was up from 14 to 18 gallons. Automatic transmission filler, door and hood latches were redone – the last two with safety in mind. Dart showed Chrysler's love of the torsion bar. It used one to keep the hood open, not a cheap piece of wire. As expected, front suspension was independent, with lateral, non-parallel control arms, torsion bars, and tube shocks. The live axle was sprung with four leaf springs on 170in^3 Darts, five on 225 Slant-6's, with tube shocks damping out the ride on both.

Refinement ran to tailored wall-to-wall carpeting with polyethylene backing, and push-button operation for the TorqueFlite auto. The aluminum 225in^3 diecast block option ended early in '63 MY, and 1964 would mark the demise of push-button automatic on Dart. However, on debut, Dart boasted a new, silenced air cleaner, redesigned starter motor for better cold weather operation, improved alternator, larger axle bearings, and easier park brake – all spoken of with reference to improvements over Lancer.

Such utilitarian discussion was done partly because of the post-recession climate, and also due to the rise of the VW Bug as economy car extraordinaire. Dart beat Bug soundly on scat. As a three-speed manual with standard 3.23 rear axle, *Car Life*'s March '63-tested Dart 270 four-door sedan did 0-60mph in 13.3 seconds, and the quarter-mile in 19.3 seconds at 72mph. This Dart could hit 95mph, and achieve fuel economy in the 17-20mpg range on regular – all sound reasons to consider Dart instead of the smaller imports.

By this stage, Chrysler weren't offering Hyper Pak 225s, but the aftermarket could lend a hand. Offenhauser and Weiand did intake manifolds to seat a 4bbl carb. Iskenderian did a range of sports camshafts, and there were companies that could supply high-compression pistons. The most exciting rear axle ratio was 3.55:1. The four wheel drums were adequate for the performance, and measured 9x2.25in front and 9x2in rear. Sensible optioning would see a TorqueFlite autobox and power steering: the former had no effect on gas mileage versus the manual box, and the latter cut lock to lock turns from 5.3 to 3.5. There was Chrysler Corp's limited power steering feel, but it was a necessary option.

Instrumentation was complete for the era. From the left, Chrysler's asymmetric style saw a large 4in diameter speedo, then small rectangular gauges for gas, coolant temp, and battery charge. There was an oversize low-oil warning lamp above the steering column, with turn signal lamps on either side. All this, and 5.4ft^3 more trunk space than Lancer, made for an ideal family car. Many folks in North America agreed, as did some outside. Darts were made at factories in Belvidere, Hamtramck, Los Angeles, St. Louis, plus Argentina and Madrid, Spain; a 'world car' before anyone used the expression.

Dart excitement – V8s, Rallye suspension & ... radials?

With senior compact Dart a sales success, detail improvements were worked-in for 1964. Front parking light decoration, an aluminum grille band at the front, and etched, radial, bright outer-bezels at the rear. GT's had their rear fascia script replaced with the triangle GT badge. There was a new side molding trim style, 3in wider rear window, and larger round taillights. HVAC controls were placed higher, and better window seals saw less exposed black rubber, and more glass for visibility.

There was a new optional electric windshield washer. It was activated by a button in the wiper's twist knob. The optional power tailgate button would now only work with the ignition on, to keep kids safe. Chrysler's stronger door latches were incorporated. They were developed in crash tests at its proving grounds. For the optional a/c, a narrower evaporator/blower box increased footwell space, and the aluminum condenser saved weight. More exciting was the arrival of Chrysler's new four-on-the-floor option, and Sure-Grip lsd. These two items went with the new V8 option like lox and cream cheese!

As stated by *Motor Trend*'s Jim Wright, in February 1963, a sporty car like Dart needed a performance option. As a sign of the times, he noted not everyone was into economy. As per Valiant, the mid '64 MY 273 small block V8 was a derivation of the 318. Engineering changes were directed at less weight, cost, and ease of compartment fit. The figures were 180bhp at 4200rpm and 260lb/ft at 1600rpm. The result was over a second knocked off the 'ol ET versus 225 Slant-6. Chrysler engineers made many engineering changes to keep the A-body in balance with the 273, but didn't touch the brakes.

The 1964 A-body was still limited to 13in rims, so no braking system changes,

Dart revisions for '64 MY included front fascia parking lights, and horizontal aluminum grille band. This was the last year for the Dart push-button automatic. (Courtesy Detroit Gearbox LLC. detroitgearbox.com)

A 1964 Dart GT convertible. The public really dug senior compact Dart. Sales were running nearly 40 per cent ahead of 1963's stellar result, even before the mid '64 V8 option! (Courtesy Sjoerd Sonsma)

and a lot of fade after the first stop. However, the Mopar magicians had the situation covered, with three major changes for 1965. Dart was getting very sporty; no push-button TorqueFlite, but a new hi po 235 horse 273 V8 option arrived. The motor cost $99.40, made max power at 5200rpm, and 280lb/ft at 4000rpm; high rpm for a family car, but the solid lifter format made it durable. This engine could also be found in Valiant and Barracuda.

Car and Driver tested a '65 Dart GT two-door hardtop with TorqueFlite. They got 0-60mph in 8.2 seconds, 16.9 seconds at 87mph in the quarter-mile, and estimated top speed at 120mph. Fuel economy for the 3040lb bolide was stated at 12-18mpg, but no-one cared about that in 1965! Of more interest was Dart's new four-slat grille trim, with five vertical dividers. Taillights went to recessed ovals, joined by a horizontally ribbed trim piece, with no GT signage. There were fender/door lower triple faux exhaust ports, and rear fender 'DART' callout in chrome script.

Whether hardtop or ragtop, Dart's GT package involved buckets. According to *Motor Trend*'s Jim Wright, Dart's buckets were unusual for a domestic, in that they actually provided support! (Courtesy Sjoerd Sonsma)

This 1964 Dart GT 225 convertible originally hailed from California. It was brought to the Netherlands in 1998. For '64 MY, the 145 horse 225-6 had its CR raised from 8.2 to 8.4:1. (Courtesy Sjoerd Sonsma)

Dress-up underhood items saw crackle black and chrome finish head bolt display for the valve covers. On the outside a low-restriction air cleaner, special intake manifold, and low-restriction single exhaust system that attracted cops, not girls! As concurrently seen on Barracuda Formula S, the four barrel carb had 1.437in primaries and 1.562 secondaries, plus 248-degree duration cam with 26 degrees of overlap. The 180 horse 273 V8 had 240 degree cam and 16 degrees of overlap. Valve lift increased from 0.395in to 0.425in on the intake side, and was 0.405in on the exhaust. There were stronger valve springs, but valve sizes stayed 1.78in and 1.50in for intake and exhaust respectively.

Domed aluminum pistons, with steel inserts raised CR to 10.5:1, and said pistons had lighter pins. With all this sportiness afoot, Dart's optional console arrived in '65 MY, along with better brakes. In 1963, it was optional self adjusting drums and, in '64 MY, power drums. The new '65 MY option involved front disks – the four piston caliper Kelsey Hayes system, seen on Mustang. This $54 option came with 11.125in cast iron vented rotors. The sentiment was that this was the largest type of car one could drive with non-power disks.

The big disks were possible, because of the newly available 14x5.5in HD steel rims. Associated with them was a special handling/tire package. According to *Car Life* in October 1965, it all created the second most roadable domestic car, after Sting Ray Corvette. The Rallye suspension package was an A-body phenomenon, which included front 0.82in swaybar and larger diameter 0.87in torsion bars. Front ride rate was firmer, rising from 105 to 115lb/in. The rear now had six leaf springs, with ride rate increasing from 120 to 150lb/in. Shock absorbers were re-valved stiffer, and what about those Goodyear Blue Streaks?!

Goodyear Blue Streaks were bias belted four ply tires, with a six ply rated carcass. The wide, square-shouldered design took lessons from racing. It had the biggest footprint and lowest cord angle of any OEM tire, at 29 degrees. The cords were made of nylon. Similar competent results to Barracuda were achieved. Dart's overall handling was helped by being a 20 per cent more rigid unibody than Lancer. There were fewer joints and seams, one piece stamped fenders and more substantial B-pillars. However, it all seemed to go better on radials.

Radials weren't an official option, but leading American tire makers had been sending their designs to the Big 4 for testing. The domestics were tardy in adopting disk brakes and radials. Radials did deliver better wet road adhesion, tread mileage, and ride comfort – on Chrysler Corp cars with power steering they even improved steering feel. However, improved ride comfort would only come if the suspension was tuned softer to suit radials. No domestic automaker was doing this in 1965, however, radials made by American companies for American cars were around at this time, albeit for compacts.

The biggest radial one could find was Firestone's new F-100 6.50-14. Firestone released them in the spring of 1965. In modern metric speak they had a 165mm footprint, versus Blue Streak 6.95-14's 175mm equivalent. Combined with the affordable, popular, and practical optional trailer towing packages of the day, and lsd, they firmed up domestic compacts into demon handlers, if the road was smooth.

Early emissions control

Yup, the F-100s were smaller than Blue Streaks, but placed on a '65 Dart GT hardtop with 235 horse 273 V8, lsd, front disks, and Rallye Suspension, the empirical results favored radials. *Car Life*'s October 1965 report on just such a Dart, revealed the difference on HD 14x5.5in rims. Dart GT cornered 10mph faster, and could use all its power on the F-100s. It also acquired tidier oversteer.

Chrysler Corp were also ahead of the curve on pollution controls. Closed crankcase ventilation systems were mandatory in California and New York. The domestics were soon putting this system on cars for all states, but Chrysler had something more, called a 'Clean Air Package.' The Clean Air Package even exceeded the strict dictates of the Golden State's smog Nazis. It involved leaner carb settings, recurved distributor, limiting choke duration, and a spring-loaded vacuum-sensing valve. The valve controlled spark advance on deceleration, and retarded spark advance at idle. Said valve was placed on the rocker cover, with a vacuum tube between carb and distributor. Ignition timing adjustments would be based on vacuum changes. The idea, which would be explored later, when emissions got serious, was to make an engine run hotter and burn the air/fuel mix more thoroughly.

Chrysler's chemical laboratories assistant chief engineer, Charles M Heinen, submitted a paper to the Society of Automotive Engineers (SAE) discussing the Clean Air Package, and the need for more frequent tune-ups to keep the 225 I6 at 14.2:1 air/fuel ratio, and hence clean. Tune-up specs were placed on a plate, next to the sensing valve. Joint availability of vacuum-sensing emissions control, and radials, would have to wait until 1974! The former was in the guise of Chrysler's Orifice Spark Advance Control Valve (OSAC).

Jay Howell's Dart Charger of 1965-66 became the Polka-Dart under Howell/Dick Branstner stewardship. It was a moderate success. (Courtesy Don Garlits, garlits.com)

1965 Dart Charger

1965 saw Dodge offer a slightly more sedate Dart, with newly-available hi po 273 LA small block. It was called Dart Charger, and was a yellow Dart GT hardtop with black interior, Charger badges, cosmetic flourishes, and the best street options, joining Commando 273's 235 horse solid lifter V8. It was the first Dodge with the Charger moniker, and lasted one year. In 1966, the iconic name went to the B-body. For Dart, the 'Charger' title remained only as a 225 Slant-6 marketing tag.

Dart goes racing

Dodge's and Chrysler's success continued. The former was the first to better its 1963 result in the sales rat race. Dart developments for 1966 were slight. Taillights were less recessed, larger, and more flush with the horizontal chrome surround. The venetian blind front grille had two prominent vertical sections, down from a former five. 1965's faux profile exhaust ports were history, and 1966 would be Dart wagon's final year.

Under Dodge-sponsor influence, Frank Wylie told Don Garlits not to race 392 V8s any more. Garlits took over the Polka-Dart, put in a double disc dragster clutch, repainted the car basic black, and rechristened the machine 'Garlits' Dart.' (Courtesy Don Garlits, garlits.com)

At a time when a normal Dart had a 111in wheelbase, the new wave of funny cars represented the evolution of drag racing. The altered wheelbases of A/FX were combined with roadster fiberglass bodies. (Courtesy Don Garlits, garlits.com)

You couldn't option $316.20 a/c on a 170in^3 Dart, but this didn't trouble Don 'Big Daddy' Garlits. Garlits' Dart didn't have a/c, probably because eventually it didn't even have a roof! These items have less importance in drag racing.

Ford was putting the heat on Mopar, in terms of the Thunderbolt and later, Mustang. Chrysler wanted a super new racing V8, and smaller car to challenge Henry's pony. The answer was Hemi, Dart, and Barracuda. Dodge sponsored Don Garlits at the time, and Dodge's PR guru Frank Wylie intimated to Big Daddy, that Chrysler wanted an exhibition stocker. Time was of the essence, given Mustang, and the relative failure of drag racing Dodge Charger's launch. That car's initial promotion hadn't gone well, with a loss of over $100 grand!

Don Garlits had been running 392 V8s, and experimenting with blown 426 Hemis. Dodge sent him a new '65 Dart in the fall of 1964. At his Florida workshop, Don Garlits and Connie Swingle stripped said Dart. They put a Hemi with custom parts in the back seat, for improved weight distribution and traction. This fuel-burning Dart had a Dean Moon hydraulic throttle, and B&M Torkmaster transmission. Chrysler sent down a film crew to capture the Dart in flight. Its maiden run would prove to be its last.

Connie Swingle pleaded, "Gar, don't hurt my toy, now." Unfortunately, Don Garlits employed his usual full throttle style. Said throttle jammed wide open, and Hemi Dart broke in two when Garlits slammed on and locked the front brakes. The torque converter then exploded, and the subsequent burst of flames singed Don Garlits, as he quickly jumped out of the Hemi Dart. Garlits was wearing a T-shirt and no helmet!

At around the same time, Jay Howell campaigned a similar Dart Charger, named after the special 1965 yellow Dart GT hardtop. This racing car became the Polka-Dart, and achieved some success. Meanwhile, Frank Wylie told Don Garlits to stop using 392s, and switch to Hemi V8s in his dragsters. Wylie also gave the instruction, as part of Dodge sponsorship, that Garlits takeover the Howell/Dick Branstner Polka-Dart. Don Garlits did this, and put in a double disc dragster clutch. The Polka-Dart was painted basic black, and fittingly rechristened 'Garlits' Dart'.

Unfortunately, Garlits Dart was wrecked by Emery Cook at Detroit Dragway, but the concept's evolution continued. Garlits had finally gotten the Hemi to perform, by advancing the V8's ignition timing. Once again with impetus from Frank Wylie, a Garlits Chassis Special plus rollcage, and fiberglass body sans top, was combined with a blown Hemi 426. This tube-chassis Don Garlits Dart follow-up became the first funny car to hit 200mph! This 1966 achievement, was enshrined with the headline: "Dart Goes 200." A funny car is a dragster with a plastic body. Don Garlits paved the way for machines like the Charlie Allen 'Crenshaw Dodge'. This blown '68 Hemi Dart 426 had a 115lb fiberglass body over a tubular frame weighing 125lb, plus just about the complete contents of a Mickey Thompson catalog. The body was by Contemporary Fiberglass of Azusa, California.

Don Garlits was in favor of funny cars, and anything that brought spectators to the sport of drag racing. Unfortunately the NHRA didn't share his enthusiasm. They didn't immediately accept funny cars, and chose not to permit roadsters in experimental stock. So, by late 1966, Garlits was back to

NEW ULTRA STOCK CHALLENGER -- Moving into ultra stock (funny car) dragracing competition with a vengeance, Don Garlits, king of the fuel dragsters, has fashioned a 426 Dodge Hemi-powered roadster-type '66 Dart. Emery Cook, 39-year-old dragster vet, shown here with the altered ultra stock, will handle the driving chores. Supercharged and fuel-injected, the new Don Garlits' Dart has an 18-foot body on a 120-inch wheelbase. Cook expects to attain e.t.'s in the low 8's at speeds above 180 mph.

From: Don Garlits, Inc., 85 Minnesota, Troy, Mich. (66-3372) Lov

50

"Dart Goes 200" said the newspaper. In 1966, Don Garlits' machine became the first funny car to hit 200mph. Don Garlits has always been pro funny cars – anything to get folks interested in drag racing. (Courtesy Don Garlits, garlits.com)

traditional dragsters, and, like Frank Wylie had suggested, put a Hemi in the rail.

Dandy Dick Landy also had some use for the '66 Dart in A/FX. The altered wheelbase category found a home in the AHRA. To stay competitive, Landy had downsized from his full-size '65 Dodge, to said Dart. A light, sophisticated chassis was combined with a custom Landy-built 426 Hemi. The motor employed Hilborn mechanical fuel-injection, a long ram tube intake manifold design, and ran on a nitro/alcohol mix. It proved successful, until the Logghe tube chassis SOHC 427 Mercury Comet Cyclone blew in.

Chrysler also supported a one-year SCCA foray with Dart, with Group 44 homologation hardware for Dart 270 and

According to the order forms, it was theoretically possible to specify the solid lifter 426 Hemi V8 in the Dodge A-100 van! (Courtesy Craig Lundquist, round2corp.com)

1968 GTS sans decals was a real street sleeper. The basic Dart shape, was the antithesis of the broad, C-pillar shouldered muscle car era. Like *Car and Driver* said, it took the engineers to make GTS standout. (Courtesy andy440.com)

A GTS could have a 340 or a big block, but didn't carry an exterior displacement callout. However, in outright performance, 340 versus 383 V8s were a wash at mid-14s. (Courtesy Ken Bowser, andy440.com)

GT hardtops. Bob Tullius, and Quaker State sponsorship, took a 4bbl 273in^3 Dart to victory in the 24 hour Marlboro Motor Raceway enduro, with Tony Adamowicz coming fourth. Dart 273 V8 also won at Riverside; plus, Ron Grable garnered a SCCA A/S title in 1966, thanks to his trusty '66 Dart two-door hardtop.

Even though a dearth of big blocks had largely kept Dart out of Super Stock, drag racing was where the action lay in this era. To qualify for NHRA D/stock, Dodge offered racers the 1966 D-Dart. Fitted with a very hi po 273, 275 horse D-Darts had a larger carb and sports cam. An 8.75in rear differential, and 4.89 rear gears, were further hardware upgrades outlined in Chrysler dealer bulletins concerning the sale of D-Darts.

Dart's 1967 makeover

1967 saw a complete revision of the Dart line; mechanical and cosmetic upgrades, and no more wagon. It heralded an era when Dart commercial and quarter-mile performance went into high-gear. Styling became even more squared up; very fashionable, at the time. The concave grille insert had fine vertical slots, and horizontal rectangular parking lights. The curved side glass was a first on Chrysler compacts, and the backlight featured compound inverse curves. The latter created a formal look. This was complemented by inward facing, vertical stack taillights, with mid amber segment turn-signal flashers.

The A-body's steering components were changed, and larger engines could be fitted. There were wider front track and frame rail spacing. New K-members made a big block possible. However, safety was king, mainly by federal decree. The dual circuit braking system had a dash telltale lamp. Along for the ride were collapsible steering column, dash/sunvisor padding and front shoulder belt anchors for outboard positions. 1967 engines saw the familiar 225 I6 and 273 V8s continue, the latter still available in 10.5:1, dome piston, 235 horse form.

The perfect family car? *Car Life* thought so. In its 1967 July issue, it was said: "This is all the car anyone with good sense would ever need." The test example was a low-19s four-door sedan with 225/TorqueFlite column shift, a/c and stock 9in drums. For the family guy looking for thrifty transport, this $3118 steal with power steering really delivered in 1967. And what about that $55.70 Trailer Towing Package? These packs were an affordable means to get durability and handling benefits on all

1968 GTS: the optional $51.10 console-mounted tach was unreadable, but the high performances of 340/383 V8s were unbelievable! (Courtesy andy440.com)

domestics. In Dart's case, a 20qt radiator with special air seals, 17in diameter six blade fan and steel shroud, 3.23 rear axle, HD turn signal flasher and 10in drums.

The pack's HD suspension had revised torsion bars, with front stiffness raised from 85 to 103lb/in. The rear six leaf spring rate was 130lb/in, not the 110lb/in of the stock five leaf job. A 0.88in front swaybar was included too, and all this good stuff gave controllability akin to the Rallye suspension option; it also banished bottoming out on rough roads. A base price for the mid-spec trim Dart 270 four-door was $2362, but when practically optioned it was a three grand plus ride. Dart was a sensible, fashionable senior compact that now sported a trendy strip speedometer. However, true fun lay just beyond the upscale GT.

Dart's sporty GTS

For 1968, Dart adopted a cleaner front fascia with round inboard parking lights, central vertical divider and mid-horizontal band. There was also a new chrome character profile trim. Federal Safety Standard (FSS) 108 required sidemarkers, separate seatbelt shoulder harness and non-glare matte finish on windshield wiper arms. For emissions, Chrysler Corp made its Clean Air package a 50 state standard fit. Keeping Dart practical were revised windshield and backlight rubber gaskets, and trim lock strips.

This '68 GTS rolls on 1970 HD 14x5.5in rims. Strangely, senior compact Mopars stayed with this rim size through the mid '70s. Like its rivals, Dodge offered optional bias belted E70-14s, but not 14x7in rims. (Courtesy Ken Bowser, andy440.com)

1968 V8 engines started with the regular fuel 273, then the new 230 horse 318 V8. However, major interest was in the hi po 273's replacement, the 340 V8. Once again, all these powerplants could be found in Barracuda. The 340 had the distinction of being the only US production engine of the hour with a forged crank. It was a hydraulic lifter V8

Hemi Dart 426 involved a solid lifter format, cross ram intake manifold, and dual 4bbl Holleys. The 80 cars were Hurst Performance-built, and dominated their class in 1968. (Courtesy Ken Bowser, andy440.com)

Above: The super stock Hemi Dart 426. Hurst Performance made the 80 LO23 Hemi Darts on Dodge's behalf, as homologation specials for drag racing. (Courtesy Ken Bowser, andy440.com)

Right: Another '68 GTS at Carlisle. More so than any other domestic compact, Dart was regarded as both an economy car and a high-performance car in equal measure. (Courtesy Ken Bowser, andy440.com)

with a big cam, so the 3.91 rear axle was necessary to stop getting bogged down on take off. The 340's main bearing alloy shells had 20 per cent tin, 1 per cent copper, and oversize intake and exhaust valves that ran to 2.02in and 1.6in, respectively. Power was 275bhp at five grand, and 340lb/ft at 3200rpm; all under-quoted, so the kids stood a chance against the insurance companies.

Car insurance for a speedy ride was like visiting a casino where no-one ever won, so it helped that the Dodge said 340 went into was a street sleeper, not a stone cold fox. Indeed, there was no 340 exterior badge or decal callout. Not all Dart GTS cars had a 340 – one could option the 300 horse 383 big block. That said, to fit the big block in the A-body hole required a more restrictive exhaust manifold, so the GTS 383 was no quicker than the 340 edition; and with the small block, a four-speed could be optioned, and gasoline could be saved – not that

Whether it was the lighter A904, or the A727 for big block V8s, the TorqueFlite slushbox was considered Detroit's best automatic. It was supplied to AMC, and several exotic car brands, too. (Courtesy Ken Bowser, andy440.com)

A convertible Dart GTS at the Carlisle car show. TV detective *Mannix* drove a British Racing Green Dart 340 convertible; the ragtop was custom built by George Barris for the 1967-68 and 1968-69 TV seasons. (Courtesy Ken Bowser, andy440.com)

the latter really mattered when gas was as cheap as air!

Dodge said the 'S' in GTS stood for Sport. It also called the car's faux hood intake trumpets, power bulges. No one was arguing when zero to 60 was six seconds, and mid-14s were a piece of cake. This car could hit 100mph one second faster than a mid '80s TPI Vette, comparing *Car and Driver* recorded times of 14.5 seconds versus 15.7 seconds. You see, there was nothing plastic about GTS. When it was stationary, sharp-eyed sports fans would notice GTS' part black-out treatment of GT's divided grille – plus 'GTS' script for the hood and tail's leading edges. However, the famous bumble-bee tail stripe was optional.

Hemi Dart LO23

When it came to the 440 and Hemi 426 powered Darts, Dodge contracted Hurst Performance. Hurst did 50 Dart GTS 440s in 1968, and over 600 in 1969. Apart from these RB block cars, there was the SS/B class directed LO23 Hemi Dart of 1968 – a homologation cousin of BO29 Barracuda. Eighty LO23s were made, with the usual non warranty disclaimer and pre-release sent from Chrysler's Auburn Hills, Michigan HQ to Dodge dealers nationwide. It stated the objective "... to sell class B super stock drag racers from factory - race ready '68 Dart GTS." The Hemi Darts were street legal, but street usage wasn't recommended. Their purpose for 'supervised acceleration trials' was explicitly stated.

It's true, LO23 and BO29 Hemi A-bodies made the '62 Ramcharger Dart 413 seem like a daily driver Prius. However, for a 1968 street machine capable of high-tens at 130mph, civility wasn't expected. The drag racing hardware and modifications were expected: fiberglass fenders and hood, lightweight acid-dipped steel doors, and thin side-window glass. There was no window winding mechanism; just a strap at the bottom of the glass, plus a strap on the inside of the door to keep the glass up. The interior had two Dodge van buckets, but no heater, console, sound deadener, radio, carpet, or armrests. No side reversing mirrors, because who cared what was back there, right?

The hood had functional hoodscoop and four securing pins. Lifted off, it revealed the 12.5:1 Hemi 426, with two Holleys on a cross ram aluminum intake manifold, solid lifter format, roller timing chain, and transistorized dual point distributor Prestolite ignition and solid core wires. The mill was joined to the four-speed A833, or the new TorqueFlite A727. With the former a Hurst linkage floorshift led to a HD clutch, steel bell housing, special torque shafts and pivots, but no synchronizers. The associated Dana 931/44 differential dealt with a

4.88 HD rear axle. For slushbox fiends it was a Hurst manual floorshift, 2600rpm high stall torque converter, leading to a Dana 831/44 differential in large stem pinion section form, plus 4.86 rear axle ratio.

Customized Darts & Dart Swinger

Keeping LO23 and BO29 in balance were HD radiator, seven blade fan, deep groove pulleys, and high capacity oil pump. The Hemi always ruled on the strip, but on the street, a Swinger 340 or GTS were more attainable auto aspirations. Like that K-tel guy said, "as seen on TV." Popular TV detective *Mannix* drove around in a British Racing Green GTS 340 convertible: a car custom prepared for the show by George Barris. It was used during the 1967-68 and 1968-69 TV seasons.

If *Mannix*'s GTS wasn't trick enough, then try Barris' Daroo I – this Dart GT based custom car had a chopped/removed roof top, and futuristic body painted Pearl Honey Yellow. Inside, a far-out center console instrument stack helped operate the 340 V8/four-speed combo. Then there was Daroo II, with Dodge collaborating with Ron Perau's Imperial Kustoms of Tulsa, Oklahoma. This open custom car had 'Contour' seats, a 'Frosted Fire' paint job involving Arctic Orange and Pearlescent Gold, and stock 340 V8/TorqueFlite power team.

Also from 1968, something for the little lady, namely the 'New Dimension' Dodge Mini Dress costing $3.75. Available in various other styles like Terry Skimmer at $7.75 or Acrylan A-Line for $10.75, they consisted of three horizontal bands, interrupted by the wearer's vital statistics. According to Dodge's Jack McFarland, the company had received interest from around the world. This included San Juan, Australia and a request letter from West Germany, asking to be European distributor. The New Dimension Dodge Mini Dress – because it pays to advertise!

The GTS was a member of the 'Scat Pack' high-performance range, so it could have a bumble-bee stripe. It could also have Sure-Grip lsd, front disk brakes, console, and vinyl roof; the last two for $50.10 and $75.10, respectively. However, these were all options. The base Dart GTS cost $3189 and included 340 V8, four-speed stick with introduced Hurst shift linkage, HD suspension, profile stripe, and E70-14 tires. The Hurst four-speed linkage had been absent during 1966 to mid '68 MY, when GTS was introduced. Chrysler Corp had been trying a vague reverse lock out 'wet noodle' shifter in the interim. With Hurst's return came an under-dash warning light informing the driver, by ESP no doubt, that they were in reverse.

In reality, GTS was only moving forward. In the October 25, 1968 issue of *Autosport*, Simon Taylor recounted

A 1969 Dodge Dart two-door hardtop: such a hot seller, Dodge could afford just slight styling revisions for '68 and '69 MYs. (Courtesy Greg Ondayko)

how his GTS four-speed smoked a miffed XKE owner, whose honey in-tow was obviously much hotter than his Jag. The GTS on test didn't have that bumble-bee stripe, disks, or lsd, but it did get the job done on 340 V8 power. This was probably why kids didn't waste their loot on fancy options. This thinking paved the way for the 1969 Swinger. Swinger was Dodge's new two-door hardtop youth model. It embodied the 'less is more' philosophy during the summer of love.

The Swinger did that with a front fender decal, not a badge. In a rationalized range it was the default two-door model, starting as a stripper 225-6 for $2511. But baby, if that was the bottom, Swinger 340 was the top! 275 horse V8, four-speed, duals, HD Rallye Suspension, D70-14s, performance hood with faux intake trumpets, 340 callout, bumble-bee stripe, buckets, and an exclusive bright-red exterior hue was new for '69 MY. Swinger 340 also procured one Scat Pack membership. However, there were still a few reasons to go with GTS. The range topper could be a convertible. A big block came in GTS 383, and rarer '69-introduced GTS 440. Like the 'Cuda 440, the larger motor meant no power steering, power brakes, or a/c ... that is a real muscle car!

As for down to earth matters, from January 1, 1969, federal law required front head restraints. '69 Darts changed from the 1968 round side-markers to rectangular items, and 225in^3 cars adopted the Canadian-spec carb anti-icing system. The drum brakes auto-adjustment mechanism was also modified.

This Dart has been retrofitted with Chrysler's high output Hyperpak 225 Slant-6. Chrysler engineers did a lot of early research on using tuned intake runners to boost I6 and V8 performance. (Courtesy Jerry Heasley)

Dart had simplified its inward facing taillights, from three segments to two. The front grille was now less elaborate. A horizontal divider band, and rectangular parking lights, were all similar to the front visage of the later 1971 Dodge B-series van. 1968 Darts didn't start or run so well, given that year's emissions law dictates. 1969 brought a revised choke system. A tag off the turn signal lever instructed one to depress the gas pedal to the floor before a cold start, with half throttle warm starts thereafter. It worked well, but some issues had arisen.

Since 1966, complaints concerning braking axle tramp were getting louder; the location of the live axle caused a severe judder in panic braking, as the rear tires intermittently lost contact with the road. This reduced braking effectiveness, and often saw the tail slew sideways.

This Dodge Dart upholds Mopar drag racing tradition. By the late '60s, Hemi Dart funny cars, like Charlie Allen's Crenshaw Dodge, were tearing up the strip! (Courtesy Jerry Heasley)

This 1969 Dart two-door hardtop was featured in the April 2009 issue of *Mopar Muscle* magazine. (Courtesy Greg Ondayko)

Concerning Mopar's super light power steering, *Car and Driver*'s September 1968 GTS report suggested the level of steering feel depended on the particular car coming off the line. In any case, Mopar cars had that rare combination of being stars of road and track. The street GTS could blow away the pricey Camaro Z/28, 375 horse Chevy II, and even the Road Runner 383!

Dart downsizes for '70

More simplification for 1970s Dart arrived. No big block, no ragtop, and no GTS! Product planners wanted the new E-body to have a real pony car shot at success. So Dart decided to just be a lively Swinger, and rationalized to four-door sedan and two-door hardtop, with base, Custom, and Swinger trim levels. The four-door could be had in all three trim levels, though four-door Swinger was very rare.

Big things for '70 were the newly-standard 14in rims, and adoption of the full-size look, the latter with a mere 0.8in length increase. At the front was a longer hood with split grille, and a longer-look sloping tail was the butt of no joke. Integrated horizontal taillights now lived within the rear chrome bumper.

Still on a 111in wheelbase, and with hardtop dimensions of 196.2in length, 54.2in height and 69.6in width, new high-back bucket seats were available, with integrated head restraints. There was a new, three-speed, all-synchromesh floorshift stick. This manual transmission had constant mesh gears, strut-type synchronizers, and slotless synch rings. This three-speed could be teamed with the new 198-6. However, the top 340 V8 and four-speed combo was reserved for Swinger 340, and brought membership to the Scat Pack.

With the popular Swinger making up 30 per cent of Dodge's 1969 business, it was natural to give Swinger 340 top

As shown by this 1969 Western Sport Special (WSS) Dart hardtop, regional dealers tried to distinguish their Darts, and steer prospects to their neck of the woods. (Courtesy Larry Williams)

In 1969, plain Jane two-door Swinger hardtop garnered 30 per cent of Dart sales. This made the lavishly specified WSS a Dart of sharp distinction. (Courtesy Larry Williams)

This 1969 WSS Dart sports the trusty 225 I6, even though most buyers of the hot-selling Swinger optioned the $30.05 318 V8. This lavish Dart was a premonition of the late '70s move to downsized, smaller-engine luxo cars. (Courtesy Larry Williams)

honors. No GTS nor Duster bodystyle in 1970, but Dart still accounted for 15 per cent of the million unit economy/compact car market. Loaded Swinger 340 could have functional dual hoodscoops, with said hood painted black and secured by tie-down pins. Front power disks, Rallye suspension, 3.91 rear axle, Bumble-bee tailstripe, E70-14s on 14x5.5in rims, power steering, tach and your choice of black or white vinyl roof. All announced with standard 340 callouts.

Base sticker Swinger 340 items covered Scat Pack decals, bumble-bee stripes, 340 badging, dual scoops, Rallye suspension with E70-14s and new three-speed stick. All the other good stuff cost extra. Option extroverts could have a pedal dress-up pack, front and rear bumper guards, rim blow steering wheel and door edge protectors. The four-grand economy car had arrived! *Car Craft* tried a well-optioned Swinger 340 with three-speed stick in May 1970. It found the new 'three-by-the-knee' tricky on the strip, but still got a 14.7 second ET at 96.84mph. Mopar's four-speed was never user friendly. That's why they call 'em muscle cars! Four-on-the-floor cost $200 more; with insurance premiums being what they were, many dug the wine bottle, but couldn't afford the contents.

This early production WSS Dart is loaded with factory and dealer options. These include power steering, power brakes, a/c, map light, glovebox light, trunk light, woodgrain steering wheel, radio/8-track player (including demo tape!), deluxe carpeting with sound-deadening pack, and deluxe upholstery trim. (Courtesy Larry Williams)

This WSS Dart dealer promotional hardtop was delivered to Van Nuys' Valley Motor dealer by the Chrysler district sales manager. Dealer alloy wheel covers, door edge guards, remote driver-side reversing mirror, deluxe rocker panel trim/belt line molding and bumper guards; they all make this Dart a Mopar to the max! (Courtesy Larry Williams)

In 1971-72 Dodge Dart's performance model switched from Swinger 340 to Demon 340. A 1972 Demon 340 with automatic and factory a/c is shown. (Courtesy Carrie Leetham, hanksters.com)

The demonic Dart

In increasingly style conscious times buyers wanted pony/muscle car looks, but wished to go easy on the options. In 1971, Plymouth lost exclusive rights to the A-body coupe, and Duster got a demonic sibling. However, Dart Demon was nearly called 'Beaver' – management being clueless as ever when it came to nameplate selection.

Demon took over from Swinger, as the new youth model. Holy sport-compact, Batman! Those Mopar marvels worked in a cute demon cartoon-logo decal, complete with pitchfork, for the rear fascia and front fender. Demon also distinguished itself by adopting a Dodge style split-grille front fascia. Plus, stylish six-slit taillights.

1971 also saw the arrival of the iconic optional 'Tuff' steering wheel. Many considered it an ergonomic and visual treat, but some didn't. Swinger continued in two-door hardtop guise, but was limited to the 230bhp 318 V8. Scat Pack membership was denied. Said hardtop and four-door sedan featured quad rectangular taillights for '71 MY, and the Custom trim level continued. However, for enthusiasts visiting one of Dodge's 3300 dealers in 1971, Demon was their destination. One could order a base Demon three-speed with 125bhp 198 I6 and 6.45-14s, but the Demon 340 package was where excitement lived. The 340 was the usual 275 horse small block, with 340lb/ft at 3200rpm and 10.5:1 CR. Even in 1971, one could order the triple deuce carb option, from 'Cuda 340 AAR SCCA exploits.

In Demon 340's realm were standard Rallye suspension, with special torsion bars, HD leaf springs and two swaybars, plus oversize shocks and HD drum brakes. The last item sized 10x2¼in front and 10x1¾in rear, power brakes were

Plymouth had exclusive rights to the new coupe body in 1970. Now it was time for Duster to have a Demonic twin – much to the chagrin of conservative Christian groups! (Courtesy Carrie Leetham, hanksters.com)

Roomy for its class, the A-body coupe even had okay rear headroom. Mopar a/c was powerful, but Tuff tiller was optional. Aside from Javelin, pony cars didn't offer reclining buckets. Some say it was to curb hanky panky! (Courtesy Carrie Leetham, hanksters.com)

optional. The Demon 340 came with E70-14s, but the oversize JJ 14x5.5in rims were optional.

Three-speed all synchro stick was standard, and Demon's side stripes and 'Demon 340' decals were a delete option. Contrary to popular belief, well optioned pony, muscle cars, and Corvettes were often purchased and looked after by middle aged male buyers of some means, sometimes as a second car. To avoid looking like a prom night racer, or a parent borrowing their kid's car because the Imperial was in the shop, they might have skipped the Demon's decals.

Hood pins were available, but more necessary were A-body's comfy $89.60 buckets, and $44.45 tach brought full instrumentation – from left to right: circular gauges with large 150mph speedo, small six grand tach and large four-segmented gauge with ammeter, gas, oil pressure and water temp. Further useful options were $10.60 variable speed windshield wipers and $27.65 lighting group. The lighting group brought below-dash lighting, map and courtesy lamp, glovebox and ashtray lights, and light for the glovebox lock. Included were 30-second lamp timer delay, illumination for the ignition keyhole, plus headlamp-on warning buzzer. To avoid fading and guarantee straight stops, axle tramp permitting, the $64 front disks were recommended.

Standard on Demon 340 was high-performance. *Road Test*'s April 1971 test brought identical quarter-mile and top speed figures as a fuel injected four-speed Vette *Car Life* tried in 1964. Both cars registered 14.6 second quarter-mile and 127mph top speed. Given that the automatic Demon was 120lb heftier, and rated at 275bhp, the Vette's 375 horse seemed like artistic license. Tony Grey and the *Road Test* staff acknowledged Demon's value for money. They agreed it was one case where ad copy was legit: "Some people today seem to be building performance cars for your kid sister. They offer a brand new stripe, but the same old six."

And there wasn't much wrong with

The 340 V8 and Demon said goodbye at the end of '72 MY. However, Demon would return for 2018. (Courtesy Carrie Leetham, hanksters.com)

the trusty Mopar 225 Slant-6. One of those could be Demon optioned for gas Ebeneezers. 1971 saw the demise of Chrysler tradition concerning left-hand threaded wheel studs on the left side; now it was right-hand threads times four. 1972's detail changes involved removed central front grille divider, and surface instead of flush mount side markers. The AM/FM radio option made a comeback, and dash instruments were made over. Performance instrumentation saw a main rectangular speedo, with a bank of small, but legible, circular gauges to the left.

Demon ran into some trouble. Christian group pressure saw curtailment of Demon logos, and abandonment of the nameplate altogether by the close of '72 MY. For '72 MY there were new fender-mounted Demon badges, sans that evil character. The rear fascia decal now just involved Dodge and Dart signage. However, some '72 Demons with side and tail stripe decals retained that lil' devil. For all the accommodations, Demon never sold as well as Duster. In keeping with Dodge's background, Demon had been promoted using the high-performance angle. In 1972, the coupe went to an optional single hoodscoop with associated blackout hood. The blackout treatment had been standard on 1971's upscale Demon Sizzler.

Re-engineering Dart for 1973-76

The Rallye rims had a restyled center cap for 1972, with light argent paint. More importantly, Dart's hi po 340 V8 was reworked for 1972-73, with an eye to unleaded gas and tightening pollution law. As per Plymouth 'Cuda, the 340 featured 360 parts to reduce cost. It also had the compression ratio drop to 8.5:1, and lower 240bhp net rating. Price was going up, but performance for 340 motorvated 1972-73 'Cuda and Dart was now in the mid-15s ballpark. A '72 455 Goat could do little better. Chrysler Corp's newly standard electronic ignition was a boon for daily driver chores. Starter motors were uprated for 1973 too.

The Chevy boys had an engine mount snafu, which led to jammed throttles and a recall. To head off that devil at the pass, Chrysler Corp modified the A-body's

Dandy Dick Landy drove a Demon in Outlaw Super Stocks. Landy's modified 426 Hemi was aboard in this stock wheelbase, 700 plus horse four-speed class. DLI (Dick Landy Incorporated) continued to work with Chrysler on R&D Mopar racing hardware, even into the no lead gas era. (Courtesy Floyd Garret, musclecarmuseum.com)

K-frame with a new spool type engine mount which limited engine roll to three degrees. 1973 also saw a change to a single piston brake caliper for the front disks. They had been four piston for the option's 1965-72 era. Plus, Darts went to intermediate B-body upper ball joints and wheel bolt pattern of 4.5in, not 4in as per previous disk brake Darts. This last quality was in common with the full-size C-body too.

Darts now used a standard 7¼in differential. The HD diff dropped from 8¾in to 8¼in; with the lower horsepower around, and big block absence, there was little need for more. Automatic cars had a 2.76 rear axle, and manuals a 3.23 ratio for starters. More safety and weight, with 5mph impact bumpers and side-impact door beams. Dart's starting weight climbed from 2930lb in 1971 to 2985lb in 1973, comparing Demon/Sport coupes. Dimensions for the sporty one between the two years went from 192.5/71.6/52.6in to 200/71.8/53in for respective length/width/height comparo.

Refinement, luxury, and practicality were coming to the fore, industry-wide. 1973's Quiet Car package brought extra sound insulation, special exhaust hangers, and quieter resonators. Then there was Dart's convertriple option for coupe: Chrysler Corp's Big 4 rivals had adopted hatchback or liftback arrangements for their compacts; Dodge and Plymouth could sport a fold down rear seat and security panel, plus manual metal sunroof, for three cars in one – namely, an economy car, convertible, and wagon! On Dart, the option included sunroof and 6.5ft^3 of carpeted

A Dodge Dart Sport 340. After that controversial Demon moniker, Dodge settled on Sport at the '73 rechristening! (Courtesy Mike Nicholl, classiccarslasvegas.com)

Unlike the Dart Sport 360 Joe Oldham met, this automatic Sport 340 is loaded with options. These include the very effective factory a/c. 1973 was the 340's swan song. (Courtesy Mike Nicholl, classiccarslasvegas.com)

load area, with the rear pews down. On Duster the sunroof was extra; it replaced a mid '71-72 folding vinyl sunroof, reminiscent of the type on yesteryear's VW Bug.

Fears over rollover roof crush testing for convertibles prompted convertriple. Emissions tightening brought a Californian 1bbl base 225, seven horses and 7lb/ft down on the 49 state's 105bhp and 185lb/ft. However, Dart's tip was still sharp, thanks to Demon's replacement, called Sport: same fella, just a different hat, and continuance of the 340 V8 option. Only this motor could have a four-speed, although 1973 Sport 340 came with three-speed floorshift, HD suspension, front swaybar, and power front disks standard. The three-speed could be Swinger hardtop optioned, in theory. Things improved for 1974, when Sport switched from the emissions-affected 340 to the revised 360 shared with 'Cuda. It was a 245 horse diamond in the rough that showed Walter P hadn't lost that Mopar feeling.

1974 – sorry, no gasoline available

The harsh realities of the first gas crunch, and federal performance car regs, were outlined by Joe Oldham in *Car*'s July 1974 issue: "Talk about your sobering effects, waiting two hours for a dollar's worth of gas can do it to you." And the Washington legislators were coming fast to make cars slow, with a clean tailpipe for 1975.

Oldham remarked George Orwell's *1984* 'Big Brother' was only ten years away, along with that anti-Big Blue, Apple Mac Superbowl commercial! Chubbs couldn't have known about that, but he did know the days of the 8-liter V8 supercar were over ... or were they?

By 1973, Chrysler's 240 horse 340 V8 had log-type exhaust manifolds and 360 heads. It also had electronic ignition, and unleaded-friendly, induction hardened exhaust valve seats. (Courtesy Mike Nicholl, classiccarslasvegas.com)

You see, you can't keep a good Dodge down – the 1990s Dodge Viper was proof of that! In the meantime, Dodge Dart Sport 360 could take on all-comers. Its V8 output nearly matched AMX's 255bhp 401, and Firebird's L75 455.

Being a smaller, lighter car allowed Dart to deliver a big block intermediate punch for modest outlay – a little over four grand's worth of inflated dead presidents. Dodge was no crook; the quarter-mile stats made rivals green with envy. Ol' Chubbs got a 14.68 second pass at 92mph out of a '74 Dart Sport 360, with 3.55 rear axle ratio. The car was an automatic, had no a/c, no buckets, and no stereo – but it did have a Tuff steering wheel. The Dart Sport also rode on radials. Standard Dart Sport 360 tire was an E70-14 bias belted job. The factory radial option brought a DR78-14, on familiar Rallye rims of 14x5.5in. This all came with a mandatory option, radial tire roadability package.

The opposition had 14x7in rims, with FR78-14s as a radial option. 14in 70 series radials were still not available for sport compact domestics from US tire makers. However, even though Dart Sport 360's radials looked a little small on the rim, they performed with aplomb. Dart delivered a ride/handling compromise uniquely good for a sport compact domestic. Said radials package brought grip, compliance, and nifty whitewall decoration; less, once again, was more.

Opponents seemed to offer less for the money. Oldham's loaded Nova SS 350 with a/c did a 15.14 second ET. A '74 Firebird with regular 455 V8 was a high-14s car; the Super Duty edition, which Chubbs dismissed as a beyond reality weirdo car, no better than the

In the '74 July issue of *Cars*, Joe Oldham said Dart Sport and Chevy Nova SS were the best value-for-money speedy cars going. (Courtesy Mike Nicholl, classiccarslasvegas.com)

low-14s. The pricey 454 Vette did mid-14s. Javelin AMX 401 was 15 seconds flat; so, too, Camaro Z/28, if one told the dealer to hold the a/c and fast glass. Chubbs opined that some mild tuning would drop Dart into the 13s. A set of headers, carb re-jet, dizzy re-curve, milled heads, and stickier rear meats could do that, with a trap speed north of 100mph, however, such a bolide probably wouldn't meet the pollution laws of California's smog Nazis.

All the hardware that speed shops used to offer were still strip legal, however on public roads they were an invitation to district court, which explained why names like Yenko, Royal Pontiac, and Grand Spaulding Dodge were spoken of in the past tense. Joe Oldham mentioned how he was able to see off a jacked-up '65 Goat at the lights in the '74 Dart Sport 360. This kind of thing happened again in 1976: by then, Dart Sport 360 was no more, but a 4bbl 360 with duals pack costing $376 could be added to a $3370 Dart Sport. In April 1976, *Car and Driver* lined up this kind of Dart against contemporary speed machines. At top speed, Dart Sport exceeded a 455in^3 Trans Am. The Dodge's 121.6mph top speed wasn't far behind the Corvette's 124.5mph.

Dart's continued sporting ability wasn't lost on Dandy Dick Landy. Landy drove a Demon with his custom Hemi 426in Outlaw Super Stocks. This 700 horsepower, four-speed combo garnered Dick Landy an AHRA Pro Stock World Championship title in 1973 and 1974. Unfortunately for Dart, 1974 brought federally-mandated rear 5mph bumpers, and Interlock; this device involved a logic sequence computer that disabled the ignition unless the driver wore a seatbelt. More welcome was upgraded cooling system capacity for Dart's optional a/c, which now used the evaporator pressure regulator valve associated with full-size Chrysler Corp cars. Non-sport Darts now had the taillights placed above, rather than within, the bumper.

Dart SE – luxury unlimited

Dart's 1973-74 grille looked awfully like its '69 counterpart; no bad thing when it came to resale value, and that solid dependability Dart was known for. 1974's Unibelt combined lap and shoulder belt portions in a single-buckle inertia reel system. Practicality; that's where the money was! If few were into Dart Sport 360 or Nova SS 350, then many wanted to sign up for a Dart SE (Special Edition) or '75 Nova LN. There was nothing limited low-volume about these rides, and salesmen didn't have to twist arms to get a John Hancock on the dotted line. By the mid '70s, people were nuts about small luxury cars – extra tinsel, and hold the V8! Industry-wide, buyers wanted downsized luxo cars. Great gas mileage was welcome too.

Dart SE was introduced in mid '74 MY, in two-door hardtop and four-door guises. High-back velour buckets, folding armrest, carpeted door panels, woodgrain instrument panel, deluxe wheelcovers (faux wires), TorqueFlite auto, and Quiet Car Package. With $406.50 a/c, radials and some essentials it was a five grand car come '75 MY, with a 225 Slant-6. The $29.50 Fuel Pacer Package helped milk every last drop of Saudi-imported goodness. It was a vacuum gauge that encouraged a light foot, using the left-side dash turn signal bulb to inform the driver when they were entering the economy zone. *Road Test* said it helped.

Dart SE eased off the line more regally than a Buick, doing the quarter-mile in 21.02 seconds at 66.71mph with 225-6/auto power team, according to *Road Test*'s May 1975 info. Fuel economy with 2.76 rear gears was 15.8mpg – only 2mpg better than a Dart Sport 360 with 3.55 rear axle. Therein lay the fallacy of EPA numbers, artificial tests, and the idea of pairing a little motor with tall gearing to yield super economy from a large luxurious body: a car with an overworked small engine is often a thirsty one. No matter, for in Dart SE, Dodge had the

The 1974 Dodge Dart SE was a small luxury car, which knew its buyers so well that advertising was superfluous! (Courtesy Dodge, dodge.com)

right car for the times. In the words of *Road Test*, "If ever there was a car that should be thankful for the 55mph limit, the Dart SE is it." Ford did sell an awful lot of Mustang IIs in 1974. Perhaps Iacocca really was fired for being too successful.

By the mid '70s, the only thing going fast was people's money. Inflation and a stagnant economy (stagflation) weren't putting a smile on anyone's dial, but Dart SE was. Dodge demographics said "... the more active 20-35 age group ..." really dug Dart Sport. A virtual tidal wave lapped onto Dart SE's shores, and *Road Test* thought they knew who the buyers were: it was the core conservative buyer group. A mini luxury car for retired folks in Florida, "... the Dart SE owner knows his car is timeless." The sensibilities continued into 1975 and 1976. A new grille to frighten Ford Granada, and catalytic converter to keep California and some high altitude areas golden. Two-door hardtops had heavy gauge steel reinforcement in the A-pillar area. This was to head-off the feared federal roof-crush test at the pass.

There was an improved-safety steering column: it had multiple slots in the column jacket. It used to be a mesh design, but now a curled column jacket over mandrel bends was floor mounted. The heater was up 14 per cent in capacity, with new third blower speed and a claimed 20 per cent greater

This 1976 Dodge Dart Sport has three clear coats added to the original color. The bumpers have been color-keyed, and the rear fascia stripe is of the 1974 variety. The added American Racing Ventura alloys are an asymmetric 15x7in front and 15x8in rear. They wear 205/70R-15s front, and rear 255/60R-15s. (Courtesy Jean-Guy Le Blanc)

This Dart Sport's original bench seat has given way to buckets. A third-party source also provided the center console with cupholders. The originally 225in^3 Sport has had its TorqueFlite upgraded to V8 type A904 unit. (Courtesy Jean-Guy Le Blanc)

airflow. The rear window defogger timing cycle was doubled to 10 minutes, and sound insulation was improved. Keeping the feds happy, by January 1 1976, Chrysler A-body was in compliance of having front disk brakes standard. 1976 also saw a foot operated park brake. This replaced the under dash T-handle, present since the A-body's major redesign of 1963. The rear view mirror was now on the windshield, rather than the roof, and front grille parking lamp lenses were now amber, in place of the former clear lenses with amber bulbs.

For federal police, Dart offered the 1976 A38 Police Package. This brought HD suspension, rear swaybar, HD rear leaf springs, firmer shocks, larger brakes with semi-metallic front pads, maximum HD cooling system, high-capacity battery, and HD alternator. Motors were the usual 225, 318 and a special 360 in non-cat 49 state form with 220 horses; this top cop had real duals, but California made do with single exhaust and a cat. Wherever the action, the bigger A727 auto and cop calibrated 120mph speedo were in service. Further police touches were high-intensity dome light, and HD wiring harness for the roof top light bar. An A-pillar spotlight and pushbars were optional.

A small volume of A38 cars were built, mainly for LA and Ventura Southern California departments. However, as with other brands, cops

preferred big cars. They sided with midsize Coronet, or full-size Dodge Monaco. Dodge had more success with its Dart Sport convertriple packages; low-volume cosmetic packs, with groovy CB handles. With a nod to surf culture, Dart Sport 'Hang 10' was painted Eggshell White. Inside were multi-colored, striped material accents for the white vinyl seats and door panels. The load area, and back of the fold-down rear bench, were trimmed in trendy orange shagpile. The dashboard and center console also had orange accents.

Externally, red and blue stripes ran from the rear quarter panels. There was also a wave-riding surfer insignia, plus a surfboard design down the hood's center. Between the taillights, Hang 10's stripe pattern turned into surfboard-shaped characters, spelling out 'Hang 10'. In mid-1974, Dart Sport 'Caravan Tan' brought a desert-color vinyl bench, with multi-colored inserts and white piping. There was a gold-colored instrument panel, dashboard, and carpet. Exterior paint offered various choices, such as Dark Moonstone, Sienna, Gold Metallic, Dark Gold Metallic, Golden Fawn or White.

For patriots during 1976's Bicentennial, there was the Dart Sport 'Spirit of '76,' featuring a white exterior, red and blue side striping, plus Old Glory motif. Many were also trying to save a buck in such inflationary times; for these

Originally, this '76 Dart Sport had a 225 Slant-6, but was converted to 360 V8 power. That power is a tricked-up 350 horses; the same as the reborn 21st century Hemi 5.7! This is a one-owner car. (Courtesy Jean-Guy Le Blanc)

folks, Chrysler Corp had the Plymouth Feather Duster and Dodge Dart Lite. These were extreme economy cars, using the coupe A-body, with a 150lb weight saving. For the first time since '60 MY, an aluminum intake manifold joined a specially calibrated lean carb, and recurved ignition. There were aluminum bumper reinforcement brackets. Such alloy brackets were also used for the hood. Trunklid bracing and inner panels also turned to aluminum.

The 225 Slant-6 had already been put on a diet for 1975. Its efficiency was boosted in 1976 with a high-flow dual biscuit cat exhaust system. It featured a 2¼in headpipe and V8 muffler. The Feather Duster and Dart Lite used a tall 2.94 rear axle, and the A833 OD four-speed with aluminum casing. The EPA rated highway mileage at a fantabulous 36mpg.

Taxis also befriended the 225 Slant-6. Between 1963 and 1976, Dart taxis were a relatively common sight. These taxis involved HD front and rear black rubber floor mats, HD seat cushion springs, full foam seating, black vinyl trim, heavy canvas backed vinyl seat covers, interior door pull assist straps, and door ajar warning lamp. HD brakes, tires, shocks, torsion bars, and rear leaf springs joined a high-output alternator, increased radiator cooling capacity, and very lean carb settings for great gas mileage.

The 1976 Chrome Yellow Taxi Package sported 95 horse and 170lb/ft. The rare California A727 RG raised deck motor had 90bhp and 165lb/ft. These taxis often carried the 999 paint code (special order paint), and had the Radio Suppresion Pack and Fuel Pacer System as options.

One thing that couldn't always be optioned was a solid reputation, and Dart had that. Dart's dependable nature was mentioned by R D McLaughlin, Chrysler's VP of car sales, as a reason for the senior compact continuing in 1976. The engines, as per 1975, were the 225, 318, and limited availability of the 360 V8, with the same '75 econo axles of 2.76:1 for six-cylinder manuals, and 3.23 performance axle for V8s. Along with more '76 HVAC upgrades, Dart continued in a semi price leader capacity, alongside its eventual replacement, the Dodge Aspen.

Such overlapping was common in the industry. It helped dealers run down discounted old stock, and provided sales security as the new unknown model was introduced. The public were very used to Dart, so no wonder dealers were a little anxious. From drag racers, to compact luxury, and taxis; Dart always had them covered.

4

Plymouth Valiant – The Economy Car King

Chrysler Corp's Valiant path

Chrysler Corp's journey to Valiant was filled with entrepreneurship and innovation. Walter P Chrysler first made his mark as president of Buick; from 1916 to 1919, he turned Buick into GM's most profitable division. Chrysler was then recruited to turn around Willys-Overland, and more than 60 years later, Jeep entered the Chrysler clan. In 1925, before all that, Mr Chrysler started his company with the Chrysler Six. In 1928, the Dodge Brothers Inc was purchased and, that same year, Chrysler Corp introduced the DeSoto brand.

Plymouth was always the 'low-cost three' family car brand, competing with Ford and Chevy. DeSoto and Dodge traded places in the first half of the 20th century. DeSoto finally ended upscale of Dodge, but below Chrysler. The marque was named after 16th century Spanish explorer Hernando DeSoto, who discovered the Mississippi River[8] ... and indeed, many were the discoveries made, and new ground broken. Moving with the cutting edge of prewar streamlining, Chryslers and DeSotos became the first American cars developed in a wind tunnel. Early work was embodied, in unitary construction, by the 1934 Chrysler Airflow.

Chrysler Corp was literally on top of the world. Celebrating corporate achievement, New York's Chrysler Building was completed in 1931. At 1047ft, it was briefly the tallest building in the world! It displayed an automotive connection, through decorative Chrysler style hub and radiator caps, plus hood ornaments. With car prototypes done in 1946, Chrysler Corp could claim to have invented the hardtop coupe. Key operated ignition was also a 1949 Chrysler innovation. At this time, Chrysler and Crosley Cars were early North American pioneers of disk brakes. In 1951, Walter P and Cadillac made life easier with hydraulic power steering. Rolls-Royce didn't have power steering, heavens, that's what they had chauffeurs for! They weren't called the idle rich for nothing. It could be said that a hardtop with power steering was a big part of postwar American driving, and Walter P put the country on that road. The convenience of a hydraulic power steering system, connected in parallel to the trackrod between chassis and steering arm, made light work of parking lots; according to the critics, a little too light.

Convenience, luxury – but the heart of the motor city was soon the V8 engine, and Chrysler had a real Duesie, worthy of Duesenberg's excellence: the 180 horse, 1951 Chrysler Firepower Hemi V8. It wasn't enough that Chrysler Corp had a modern overhead valve V8 early in the postwar era; the Mopar magicians worked in hemispherical combustion chambers too, able to engineer around a pushrod rocker arm arrangement to deliver a free-flow path between intake and exhaust sides. A centrally placed sparkplug and canted valves worked by rocker arms on parallel rocker shafts. Got Hemi? Cadillac and Lincoln didn't.

In 1951, Chrysler had the numbers in more ways than one. With respective industry totals for Chevy, Ford and Plymouth of 2,648,673, 2,087,892 and 1,193,129, Chrysler Corp were one of the Big 3. By 1953's end, Chrysler Corp claimed its highest corporate sales ever. It claimed its latest 235bhp OHV 331 Firepower Hemi V8 was the most efficient US-built engine of the day. In search for budget luxury, its new two-speed automatic transmission was called PowerFlite, with planetary arranged components, and a claim of weighing less and having fewer parts than its rivals. With a torque converter box possessing a 4.47:1 ratio, it seemed to offer the highest torque amplification in the industry. In an America where domestic brands dominated the market, any stylistic or technical advancement was ad copy gold dust.

Another Chrysler Corp quality was displayed at this time: excellent fleet

The 1960 Savoy was Plymouth's equivalent to the Dodge Dart Seneca. It competed with Chevy Biscayne and Ford Custom in the 'low price three' family car segment. (Courtesy Ben Deutschman, 1960plymouthsavoy.com)

Factory accessories added to this Savoy include clock, front swaybar, power steering, power brakes, AM radio, and Hi-Way Hi-Fi RCA 45rpm record player. (Courtesy Ben Deutschman, 1960plymouthsavoy.com)

sales. With DeSoto, the corporation catered to the largest taxi market on the globe: New York. Seven out of ten Big Apple taxis were DeSotos. However, fleet sales are a mixed blessing: you get a lot of custom, but not much money per unit. It's something Chrysler Corp experiences in the modern era with its LX platform sedans.

From fleet sales to fleet of foot, Chrysler Corp were the first in Detroit when it came to combining sports and luxury. The Letter Series cars represented the first commercially successful combination of power, luxury, and speed in a high-volume automobile that used high-performance as a selling point; a concept used decades later by AMG and BMW.

That formative example of the 'Banker's Hot Rod' was the Chrysler C-300 – a revolutionary amalgam. Basically, Virgil Exner's new styling in New Yorker two-door hardtop form, with prestige Imperial grille, lower suspension, and Hemi V8, featuring dual quads and dual exhausts. The Hemi 331 made 1955 C-300 the first domestic car to hit 300 horse, with 8.5:1 CR and solid lifter cam, 1692 were made. By 1957, the Chrysler 300C had 375bhp from its Hemi 392 V8. An extra 500 bucks brought 10:1 CR and bigger cam, for 390 wild horses. The Firepower V8 was an industry leader for power, efficiency, and durability.

Walter P had the records to prove it; namely, 196 AAA stock car standards from Bonneville in September 1953. Success in drag racing and NASCAR would follow. However it wasn't all brawn and no beauty; there were other good things in the low-cost line too. In 1955, Plymouth could offer an OHV V8, and equally modern plus pleasing styling. Designer Maury Baldwin's work was honored by the Society of Illustrators, as Plymouth gained the society's 'Most Beautiful Car of the Year' award for 1955. That same year, Plymouth had a major plant retool, to the tune of $40 million.

The 1960 Savoy featured two industry stars of the year: the new OHV 225 Slant-6 and TorqueFlite automatic, together for the first time in the low-cost field. (Courtesy Ben Deutschman, 1960plymouthsavoy.com)

It all sounded too good to be true, and the path to corporate success rarely involves smooth sailing. Chrysler Corp certainly had its ups and downs in the '50s. Styling, economic conditions, and vehicle quality were challenges to contend with. For starters, Chrysler President Kaufman T Keller had a battle royale with the styling department. Keller had some definite ideas on vehicle styling; he even branded Ford and GM shapes as 'jello-mold,' decades before anyone knew what a Taurus or Merkur was. Kaufman insisted his cars should be able to hold a milk churn, in the vertical position.

Naturally, the need to accommodate the dairy belt rather cramped the designer's style; how they must have wished that their next boss was lactose intolerant! Such avant garde styling cost Plymouth two places in the sales race in the early '50s; perhaps as much as one quarter-million sales in one year alone. In '50s America, function took a backseat to style – Chrysler Corp needed to find some; and it did. President Keller hired Virgil Exner, making him Director of Advanced Styling. His studio, Chrysler Styling, worked with Italian styling house Ghia. Jointly, they came up with dream car concepts like Dart, Norseman, and Ghia Gilda. The latter was a star of the 1955 Turin auto show. It had streamlined styling, a 140mph top speed, and tailfins.

Under Exner, Chrysler Corp introduced its 'Hundred Million-Dollar Look' for 1955. High-glamor styling in contrast to the 'milk churn cars.' This slogan was fine for medium and high-priced lines, but for low-cost family Plymouths, it was the 'Forward Look.' Marketing didn't want thrifty prospects thinking they needed a million bucks to buy a Plymouth Savoy! It all looked good, but Detroit soon found marketing comes before a fall. *Popular Mechanic*'s Jim Whipple made the observation that after WWII, the American public were starved of new designs. They only had warmed over prewar jobs. Detroit soon stepped in with a game of auto one-upmanship, only rivaled by the buying habits of suburban neighbors.[9]

The tailfin's origin is linked to GM's Harley Earl, and Virgil Exner. In Chrysler Corp's case it was the bolted-on fin of 1955. Unfortunately, all the dreaming led to the late '50s recession. Edsel wasn't Henry's number one son, Cadillac discovered Eldorado wasn't paved with gold, and the Forward Look counted for little, what with Studebaker Lark and Rambler American selling up a storm.

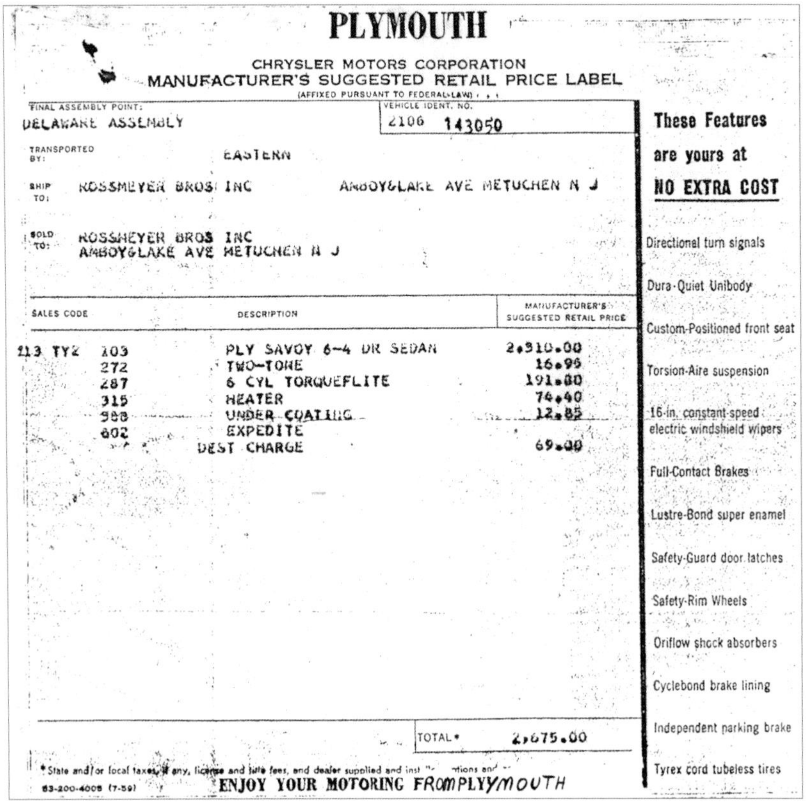
PLYMOUTH

CHRYSLER MOTORS CORPORATION
MANUFACTURER'S SUGGESTED RETAIL PRICE LABEL
(AFFIXED PURSUANT TO FEDERAL LAW)

FINAL ASSEMBLY POINT: DELAWARE ASSEMBLY
VEHICLE IDENT. NO. 2106 143050
TRANSPORTED BY: EASTERN
SHIP TO: ROSSMEYER BROS INC AMBOY&LAKE AVE METUCHEN N J
SOLD TO: ROSSMEYER BROS INC AMBOY&LAKE AVE METUCHEN N J

SALES CODE		DESCRIPTION	MANUFACTURER'S SUGGESTED RETAIL PRICE
213 TY2	103	PLY SAVOY 6-4 DR SEDAN	2,310.00
	272	TWO-TONE	16.95
	287	6 CYL TORQUEFLITE	191.80
	315	HEATER	74.40
	388	UNDER COATING	12.85
	802	EXPEDITE	
		DEST CHARGE	69.00
		TOTAL*	2,675.00

*State and/or local taxes, if any, license and title fees, and dealer supplied and installed options and ...
83-200-4005 (7-59)
ENJOY YOUR MOTORING FROM PLYMOUTH

These Features are yours at NO EXTRA COST

- Directional turn signals
- Dura-Quiet Unibody
- Custom-Positioned front seat
- Torsion-Aire suspension
- 16-in. constant-speed electric windshield wipers
- Full-Contact Brakes
- Lustre-Bond super enamel
- Safety-Guard door latches
- Safety-Rim Wheels
- Oriflow shock absorbers
- Cyclebond brake lining
- Independent parking brake
- Tyrex cord tubeless tires

This Savoy left the Newark, Delaware factory on May 26, 1960, and was delivered to Rossmeyer Chrysler Plymouth (Metuchen, NJ). It was bought new, with discount, for $2100 on June 13, 1960. (Courtesy Ben Deutschman, 1960plymouthsavoy.com)

Rambler even pushed Plymouth to fourth in 1960. Plymouth had been No 3 in 1957 and 1958, but recessionary 1958 saw industry sales drop 22 per cent. For Plymouth, the fall was 34 per cent. Being Chrysler's low-cost brand, one would have expected Plymouth to weather the recession better, however it seemed the public were economizing somewhere else.

They turned to European imports: Morris Minors, Renault Dauphines and, increasingly, the VW Beetle. Secondly, the success of Dodge's Dart sub-brand, showed a market existed between the low-cost bottom and the medium-price sectors. So, the rare stand-alone Plymouth dealers that only did low-cost were suffering. However, a medium-price Dodge outfit that now offered a lower-cost alternative could prosper. Dart was probably picking up medium-price prospects, downsizing in the wake of the '58 recession. Either way, Plymouth needed a new kind of economy car. One that spoke to the European import crowd in terms of cheaper-to-buy; smaller, so it was easier to drive; and less expensive to feed.

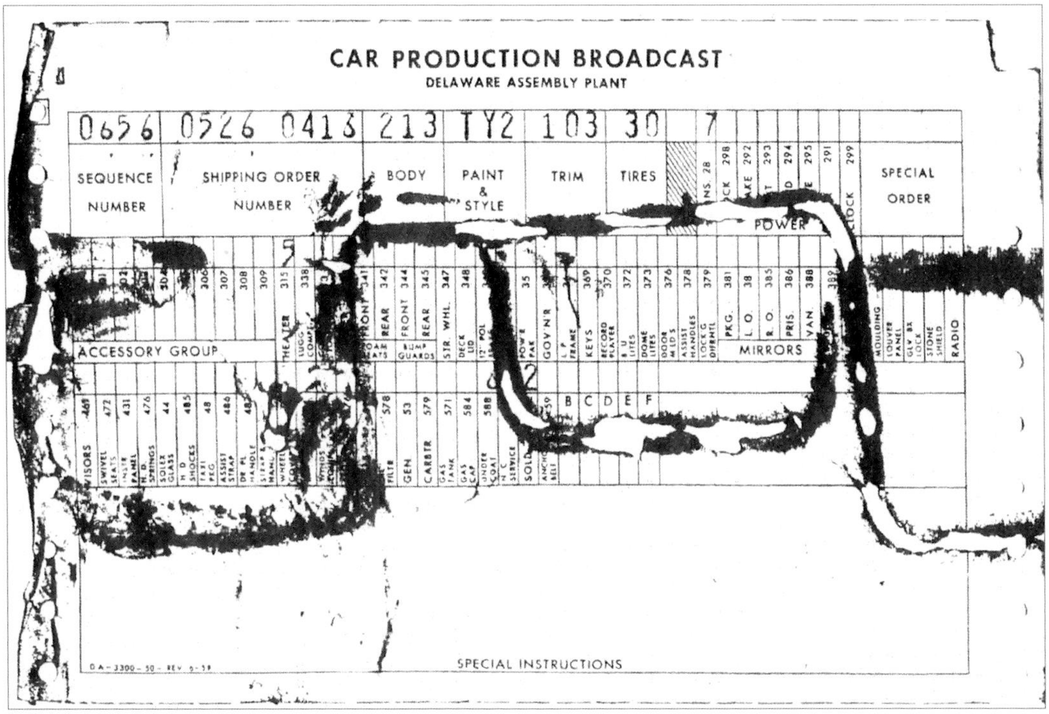

This 1960 Plymouth Savoy's build sheet was discovered behind the rear seat cushion springs. The marks on the sheet came from said springs! Nearby lived the 3.31 ratioed open differential, standard on I6 TorqueFlite cars. (Courtesy Ben Deutschman, 1960plymouthsavoy.com)

NASCAR King Richard Petty racing a Slant-6 Plymouth Hyper-Pak Valiant in the compact car support race for the 1960 Daytona 500. (Courtesy Jeff Dennison, Petty Museum, rpmuseum.com)

Chrysler Corp's Valiant economy car

Chrysler President, Lester Lum 'Tex' Colbert, was wise to the economy car problem. He recalled that, when joining Chrysler Engineering Division in 1933, an experimental small car was in the works, and that Chrysler Corp had always been working in this area behind the scenes. In the '50s, Plymouths and Dodges were offered with shorter wheelbases and, in May 1957, Tex Colbert established a committee to come up with a modern small import fighter. It was timely and necessary, given GM and Ford were working on their domestic compacts – Corvair and Falcon, respectively – for 1960, too.

Under the lead of Plymouth General Manager Harry Cheseborough, the top secret 'Project A901' was afoot at the Detroit, Midland avenue plant; 200 engineers, 20 prototypes and 57 experimental engines. On that last entity, a flat-six was in the mix. So, it seemed, GM and Chrysler were eyeing Herbie's flat-four! Immediate inspiration came from Chrysler's 1955 concept car, which was, ironically, called Falcon. This concept was styled under the gaze of Virgil Exner. It had a 276in^3 Hemi V8, two-speed PowerFlite auto, but no torsion bars. It did have a similar front fascia to what would appear on Plymouth's new compact.

The directive given to Chrysler Corp designers and engineers was that the car be truly new and different. It was under Exner's 'Forward Look' umbrella; crisp, sleek lines and a flush side look from Chrysler/Ghia D'Elegance and Adventurer concept cars, but it was unique beyond that. It was no scaled-down bigger car; quite the reverse. The look the world was about to see, would be seen on Chrysler Corp's new '62 midsize B-body. One could say it was a Valiant attempt at an American economy car. 'Valiant' was indeed its name. Henry Ford II had requested the Falcon moniker for his new 1960 compact. He did Walter P a favor.

The Hyper-Pak Slant-6 design that dominated Daytona's compact car race in 1960. So good, NASCAR cancelled the series! (Courtesy Greg Ondayko)

Falcon far better suited the Blue Oval's more conservative take on the economy car.

GM was mighty avant garde with its rear-engined, flat-six Corvair; however it was unusual, with a close eye to VW and its success. Indeed, in modern times, racier Corvairs like Yenko Stinger have been mistaken for overtly sporty Karmann Ghias. Plymouth Valiant, or plain Valiant, as it was officially known on '60 MY debut, was its own kind of thriftmobile. In fact, it wasn't immediately obvious Valiant was an economy car at all; that was the idea. Economy cars were known for being cramped, slow, and lacking in style. The comments of Chrysler President William C Newberg suggested otherwise; that is, a high-efficiency, high-performance economy car for the modern American family. Not a city runabout, but a full six-seater that could take the whole Mitch Miller gang to see Aunt Patti for the holidays!

Underlining Valiant's nature, and firing a marketing salvo at VW, Newberg added "... and it will have plenty of space for luggage in the luggage compartment at the rear." You had to feel worried for Dr. Porsche and Herbie. Valiant was thinking big. The first unibody Chrysler Corp car since the 1930s Airflow. A welded-in front subframe, with fender, quarter-panels, floor and roof all stressed for true unibody rigidity, plus a big-car ride with full-size 'Torsion-Aire' suspension, in keeping with Chrysler Corp's state of the art, unequal-length control arms and torsion bars up-front, with a leaf-sprung live axle rearwards.

Early CAD (Computer Aided Design), with IBM's assistance, was used to fast-track suspension development. Testing proposed suspension designs for load and stress. Those that passed went to the prototype stage. Computer aid was also employed to work out optimal sound proofing and reduce NVH. The body received six intensive paint spray applications. Prior to that,

For a brief time, Hyper-Pak's long intake runners caused the public to view the Slant-6 as a high-performance engine ... because it was. (Courtesy Greg Ondayko)

The Hyper-Pak intake manifold casting, recreated by Doug 'Dr Dodge' Dutra! (Courtesy Greg Ondayko)

anti-rust protection of seven chemical baths, including zinc chromate. Valiant was ready for the world. Indeed, during the car's development, Chrysler was concerned it had lost sales to GM and Ford overseas by not having an appropriate smaller family car suited to places where gas was pricey.

Valiant made its debut at the 44th International Motor Show held in London, on October 26 1959. It carried the slogan "Nobody's kid brother, this one stands on its own four tires," plus, "Engineered from the inside out," and "A sports car look all of its own;" and Valiant did it with six-cylinder inline punch. Factory figures were 95mph for top speed, and 0-50mph in 9.2 seconds for the three-speed manual. For the zombie shift edition, 90mph and 12.6 seconds respectively. Such factory data put Valiant on par with a full-size straight-eight of 1939 vintage. Plymouth's claimed 0-60mph time was 16.7 seconds, *Road & Track* recorded just 13.9 seconds, nearly equalling the time the same journal recorded for the fuel injected V8 Caddy Seville 15 years later!

Valiant's light sporting qualities

Valiant did all this with the weight-saving assistance of aluminum. Chrysler Corp was at the forefront of aluminum part die casting, with parts made at its Kokomo, Indiana foundry. Examples beyond chassis parts involved the oil pump, water pump, alternator housing, and Hyper-Pak six intake manifold. The TorqueFlite casing, torque converter housing and extension also benefitted from the light alloy. Cosmetically, even the front fascia grille and surround were aluminum. The parts weighed just 3lb, saving 10lb. Overall, 100lb was saved on Valiant thanks to aluminum, contributing to the four-door sedan with 170in^3 six and three-speed weighing just 2750lb. Valiant's dimensions were also a trim 183.7in, 70.4in, and 53.3in for respective length, width and height. Although looking larger than dimensions implied, Valiant sat on a compact 106.5in wheelbase. A true small American car.

Valiant's sporting qualities made it a racing natural. The car's racing exploits showed Chrysler Corp's engineering strength, especially in the areas of tuned intake runner length and torsion bar suspension. Such prowess was showcased in a proposed NASCAR race series for Detroit's compacts. In preparing a suitable car, Plymouth Product Planning director Jack Charipar and engineer Dick Maxwell were involved. The racing car, Valiant Hyper-Pak, was homologated via a dealer offered 'Hyper-Pak' tuning kit. The kit was available from December 1 1959, and consisted of equipment to hot up Valiant's standard 170in^3 Slant-6 to 148bhp and 153lb/ft. The newly overhead valve six, now had 10.5:1 CR, one 4bbl carb with manual choke, big cam, high-load valve springs, 15 gallon fuel tank and dual exhaust system. Enough to frighten a Jag XKE! Valiant Hyper-Pak certainly scared Falcon and

This 1961 Plymouth Valiant two-door hardtop was originally sold in Phoenix, Arizona. In 2003, it was imported from Ohio to begin a full restoration in Australia. (Courtesy Peter Flavelle)

Corvair at the compact support car race for the Daytona 500. The NASCAR support race was organised as a two part affair, held at Daytona International Speedway on January 31, 1960; Richard Petty took part.

Race one was a ten lap road course event. Valiant Hyper-Paks took the first seven places, with Marvin Panch winning. Race two was a 20 lapper on the tri-oval track. There was a multi car crash on lap four, which took out the four leading Valiants. Richard Petty's Valiant Hyper-Pak was in the melee, making his race a DNF. Fortunately, Marvin Panch upheld Plymouth honor, heading a Hyper-Pak 1-2-3, and recording an average speed of 122.282mph. Valiant became a victim of its own success: the compact race had been so one sided, NASCAR ended the series!

This Valiant was converted to right-hand drive using Chrysler's factory-designed hardware for 1961-62 Australian CKD Chrysler Valiants. This included a right-hand drive firewall. (Courtesy Peter Flavelle)

In this early '60s era, Plymouth was a drag strip natural, too. As counterparts and rivals to Dodge's 'The Ramchargers' racing hobby engineers, Plymouth had the 'Golden Commandos.' Both groups taking their names from their respective marque's top V8 monikers. The Golden Commandos were led by Ray Kobe, and were traditional hot rodders. They included dyno operators Forrest Pitcock, Troy Simenson, mechanic John Dallafior, and engine man Steve Baker. As part of the above mentioned friendly rivalry, the Golden Commandos called their Dodge counterparts 'Rumchasers,' and the Dodge guys called em' 'Commodes.' Both groups were involved in getting the new Hemi 426 into A/FX drag racing. They were also involved with the tuned intake runner development, which Hyper-Pak Slant-6 drew upon.

Valiant bodystyles and trim levels

It was a lot of sporting connection for a car released in four-door sedan and wagon forms. Made at Hamtramck; Lynch Road Assembly, Michigan; Maywood, California; Newark, and Delaware factories, Valiant was the first mass-produced car in the world with AC electrics. Thus, it had an alternator, with this innovation passed on to all Chrysler Corp cars in 1961. Domestically, Chrysler offered the cheapest American wagon. The two seating-row edition undercut Studebaker Lark and Rambler by $60. Valiant's three row wagon came with runflat Captive-Aire tires as standard equipment, implying an also standard locking cargo deck compartment. Two row wagons with Captive-Aire runflats also came with the locking cargo deck. Wagons also offered an optional insect trapping aluminum tailgate window screen.

Valiant trim levels were the basic V100 or fancy V200. The former had an exterior color choice of silver gray, blue, green, white or black. V200 added red to that palette. Inside, V100 had multi colored nylon surfaced acetate cloth, with fine metallic highlight threading and grained vinyl borders. In place of V100's rubber mats, V200 was trimmed in color-keyed carpet. Both V100 and V200 had gray rubber trunk mats. Plain V100 had a one color seat, door trim card, door molding and parcel shelf, plus metal dash. V200 jazzed up such austerity with color-keyed dash, two-tone steering wheel and bright chrome horn ring. With the exterior, V200 belied Valiant's economy car position with lots of chrome decoration.

With a January 1960 changeover, Valiant 200s made after this date were much plainer. Early cars had an 8in chrome spear on top of the front fenders, and inner reveal ring on the Sport Deck. V200 script on the dashboard, and classy stainless steel windshield and rear window frame trim, were also present. After this date, such niceties were deep sixed. Low-cost, mylar-faced plastic locking strips now framed the windshield and backlight. God bless accountants! However, all V200s possessed a continuous stainless steel molding along the tailfin. It swept down and continued to the front fender.

Quality touches saw Valiant's radiator grille bright dipped in chrome, and the central grille badge doubled as a hood release. The badge carried gold Valiant script callout against a red background. The Valiant script was repeated on the front fenders, and on the Sport Deck's center. In spite of Valiant's apparent avant garde nature, its canted tailfins, cat's eye taillamps, and Sport Deck were all Chrysler Corp fare. On the technical front, 1960 saw production improvements; namely, better oil flow to the rear two conrods, improved voltage regulator function, better cold starting and idling. Early durability problems with the intake manifold mounting studs were also addressed.

Some thoughtful mods include the addition of a later 225 Slant-6, cruise control, and front-vented disks. Apart from two new starter relays, the 22,000 miles since restoration have presented no problems. (Courtesy Peter Flavelle)

With nearly 200,000 units sold in its first year, Valiant was a hit. Given it almost took four years for Kaiser-Frazer Henry J to reach 120,000 sales, it seemed Valiant was the kind of all-purpose economy car America was looking for. In light of its overseas success, one might add it was the American car the world was looking for. However, assembly line snafus and component quality shortfalls did take some luster off for some buyers. Chrysler Corp was still a few years away from its five year/50,000 mile warranty, but at least Mr Nader wasn't knocking on Valiant's door. There was a lot of good 'ol fashioned horse sense about Detroit's new compacts. For example, Valiant wagon had 72.3ft^3 of load capacity, but

This 1961 Valiant V200 Signet was top of the original Valiant range. Marketed simply as 'Valiant' during its debut 1960 MY, the range officially became Plymouth Valiant for 1961. (Courtesy Carrie Leetham, hanksters.com)

was 2ft shorter than full-size Plymouth wagon.

Two new bodystyles arrived for 1961: a two-door sedan and hardtop. Late 1961 saw the release of the larger 225 Slant-6, and its aluminum block counterpart. Instrumentation swapped over from 1960s black gauges with white markings to white gauges with black markings. Apart from carb revisions, California's mandatory crankcase vent emissions feature became available in other states. Changes to exterior decoration saw new side trim. A 10in stainless spear, and hockey stick style lower breakline, with front fenders and door crease adorned by the stainless steel spear's cap. Tailfins were topped by three transverse chrome strips. On the Sport Deck's center was a large, round V200 plastic callout emblem, with matching items in the middle of the front fender's decorative spears.

The flagship model was the two-door hardtop V200 Signet. This was the only version keeping the central grille emblem. This ornament now had a white background, with red and blue 'V' insignia. 1961's grilles were flatter and shorter, with painted black squares. The hood release was a knob at the top of the grille's frame. The V200 Signet may have been the lowest priced US two-door hardtop with buckets, but was lavish for an economy car. Signet boasted pleated leather like seating surfaces, and custom tailored interior trim. There was deep pile carpeting, special trunklid emblem, unique headlamp frames and special side moldings.

Many were the changes for '62 MY. These included new hood and fenders, which couldn't be retrofitted to 1960-61 machines. New wraparound stainless steel-trimmed tailfins graced a rear fascia, with lower round taillamps encircled with aluminum bezels. The previously optional '61 backup lights now flanked the below-bumper license plate.

The Sport Deck spare tire stamping was deep sixed. There was now a small central ridge at the trailing edge, with V200 as a large round emblem, surrounded by rectangular Valiant script in capital letters on a black background. A similar logo rested on the front

Painted in Metallic Midnight Blue, this Valiant's aftermarket alloys have 185/65R-15s front and 245/50R-16s rear. Stock Valiants were limited to 13in rims and four wheel drums. This Valiant has four wheel disk brakes. (Courtesy Carrie Leetham, hanksters.com)

fenders. Signet had a smaller round emblem that contained the red and blue stylized 'V' logo. Valiant V200's side trim was '60 MY style, only even more so. There was a rectangular triple window effect at the rear of the body's break crease. Valiant V200 Signet front fenders were an open-centered double spear from the car's front to rear. Within lay a second body paint color.

Valiant's interior dash instruments now followed respected Plymouth big car practice: that is, the asymmetric arrangement of a large round speedometer on the left, and a bank of gauges in a top-right row. From left to right they were: gas, coolant temp, and battery charge. The left edge of the instrument binnacle had TorqueFlite's pushbuttons; the right side had the heater control buttons. Controlling the show was a shallower dished steering wheel. There were now printed circuit boards for the instrument cluster, and more fuses for the electrical system.

Further '62 MY upgrading saw a new starter motor design, and alternator, along with carb improvements. For safety in a front-end crash, formerly vertical sheer engine mounts were replaced with 45 degree ones. (In a crash, engineers want the motor to submarine, rather than crash into the firewall). With an eye to post-recessionary conditions, exhaust system material of a more rust resistant nature, and taller rear-axle ratios, came aboard. Manual steering promised to be quicker; indeed, both power and manual system steering gearboxes were newly designed. The manual steering ratio went from 20:1 to 24:1. To save weight, the manual steering gearbox now had an aluminum case, rather than the former cast iron one.

The interior is trimmed in custom dark fawn carpet. Normally, a V200 car had brocade-like upholstery matched to the exterior color scheme. Color-keyed carpets replaced the V100's rubber mats. Like full-sized models, Valiant offered a push-button automatic. (Courtesy Carrie Leetham, hanksters.com)

In late '61 MY Valiant had the 225 Slant-6 for the first time. However, the LA small block didn't appear until mid '64 MY. Even then, the meek 273 didn't have the moxie of this worked 360 V8. Black aluminum Mopar valve covers, and aluminum high-performance radiator were further custom touches. (Courtesy Carrie Leetham, hanksters.com)

Front suspension parts were virtually all new, and chassis components only needed a lube job every 32,000 miles. With all the detail engineering improvements, what of Detroit's raison d'etre – styling? It seems, even here, Valiant was achieving kudos. The praise came from the Society of Illustrators, once again. In October 1961, this respected body bestowed its 1962 Styling Award on '62 Valiant V200 Signet. The presentation was made to Virgil Exner. From trunks that could carry milk churns, there's no doubt Exner brought glamor to Chrysler Corp.

Valiant's redesign & V8 arrival

With no resting on its laurels, Valiant and Dart had all new bodies for 1963. Dart was living on a longer senior compact 111in wheelbase. For Valiant things got shorter, at 106in. Styling went contemporary with a flat plane look. Herbie the Bug was looking a little prewar! There was a gentle sweep to Valiant's new profile, plus an inverted trapezoidal front fascia grille, with fine mesh insert. The latter was Chrysler's in-house look. Engines continued with 170in^3 and 225in^3 Slant-6 power. However, the aluminum block 225in^3 option didn't last long with '63 MY Valiant. Regardless, the 170in^3 and 225in^3 motors flew on, with 101bhp at 4400rpm and 145bhp at 4000rpm respectively.

Structural improvements to increase body rigidity were employed, plus a spring-staged choke as part of the upgrades. Full bodystyle range variants covered two-door and four-door sedan, wagon, two-door hardtop and ragtop. Range-topping Valiant 200 Signet convertible could have a power top. In December 1962, Plymouth's first ever vinyl roof became a Signet option. The new '63 Valiant was a hit to the tune of 225,056 sales, and Valiant didn't stop there. The 1964 restyle involved detail changes. There was a new hood and front grille, the latter horizontal bar affair had an emphasis on width. Rearwards saw taillights, which used to live below the trunklid, now reside in the tapered ends of the quarter-panels as vertical units. Former ring style rear deck decoration, was now Valiant script placed in the right hand corner.

Valiant bumpers were now larger, and placed 2in out from the body. A four-door sedan was now 188.2in long, 70.1in wide and 53.5in high. High, wide and handsome? Well, compacts were

The redesigned Valiant co-starred as the 'Weehawken Express' getaway vehicle in comedian Jackie Mason's 1972 movie, *The Stoolie*. (Marc Cranswick)

This 1965 Valiant two-door sedan with '64 grille has been owned by the same family since new. It was purchased by its current owner from his aunt in 1977. Original power team involved 225 I6 and three on the tree. (Courtesy Rick Covalt)

In 2012 this '65 Valiant was restored, with Rob Daywalt doing the body and paint. The Plymouth has been mainly set up as a drag racer. (Courtesy Rick Covalt)

Clint Kriner of Kriner's Racing Engines built this hot Slant-6. It features 11.5:1 Wiesco pistons, .085 overbore for the 225, 0.566in lift intake, 257-degree duration cam, Mike Jeffrey race head, Clifford intake, and 600 CFM carb! (Courtesy Rick Covalt, kriners.com)

Using a TorqueFlite A904, this '65 Valiant Slant-6 has recorded a 13.82 second pass at 96mph. Plans include 4.56 rear gears for an 8¾in rear end, and new Slant-6 race motor to push this Plymouth into the 12s! (Courtesy Rick Covalt, kriners.com)

getting fancier. The four-door wagon could have V200 trim. V200 Signet implied buckets, although there were non-bucket V200 two-door hardtops and convertibles. Let no bucket get in the way of an amorous couple! As compensation, Signet buckets could be trimmed in a leather-like grained vinyl, colored black or white. The latest compact cast iron striker plate was less likely to snag clothing than the old hook-like steel striker.

For all the visual changes, A-body suspension continued with roadable, independent, non-parallel control arms and torsion bars up front, with a leaf-sprung live axle out back. Mirroring Dart, the Slant-6 had a compression ratio hike for '64. The 145 horse 225 was called the 47 buck 'Power Pak Six.' However, both 170in^3 and 225in^3 motors were one-barrel creatures that supped regular. The big news on the excitement front was Valiant's new four-speed stick option. It was Chrysler's own A833 box, giving a Hurst shift linkaged 'four on the floor,' with direct top-gear. Ratios were 3.09 first, 1.92 second, 1.40 third, and 1:1 fourth.

With Hurst shift linkage, the stick moved well through the gate, but wasn't user-friendly. Even compared to Ford and GM boxes, it was hefty, and its bigger gears made red line shifting very difficult. It was a box designed with big wedge V8 and Hemi drag racers in mind, hence the heavy-duty feel. One couldn't break an A833 box, but it might break you! All of which showed how skillful Mr Four-speed Ronnie Sox was – according to rivals, he never missed a shift. According to manufacturer release info, the four-speed was only available on Dart 225 at the start of '64 MY. However, it could be had on 170in^3 and 225in^3 Valiants. In either

This 1965 Valiant was top of the line. A Valiant Signet convertible with 273 Commando V8! (Courtesy Carrie Leetham, hanksters.com)

Valiant joined Falcon and Chevy II as compacts that offered small block V8s. The once austere economy compacts were acknowledging buyer demand for more power and features. (Courtesy Carrie Leetham, hanksters.com)

The TorqueFlite automatic's ratios were 3.09 (first), 1.92 (second), and direct top. This gave Valiant something the C2 Vette automatic lacked: a passing gear! (Courtesy Carrie Leetham, hanksters.com)

With the new '63 bodystyle, Valiant moved down one half-inch to a 106in wheelbase. Next size up in the Plymouth posse was the standard size 116in Savoy/ Belvedere/Fury line. (Courtesy Carrie Leetham, hanksters.com)

By 1964, Valiant changed to a rear right hand corner 'Valiant' script callout. It was also possible to get the factory option Sure-Grip lsd back there. However, Valiant was still at the 'value for money' forefront. (Courtesy Carrie Leetham, hanksters.com)

case, four-speed was combinable with 2.93 or 3.55 rear gears.

In sporty situations, one needs traction aid. Just as well Sure-Grip arrived on A-body for '64 MY. Sure-Grip and four-speed hinted at the next move – a V8 option! Yes, from mid '64 MY Valiant joined the V8 compact crowd. Chevy II had the 283 V8, and Ford Falcon Sprint 260 V8 had arrived in mid '63 MY. Studebaker Lark had the hottest V8s in compact land, but Rambler American and Corvair were exclusively six swingers. However, Corvair did offer Monza and turbo packs, albeit with little commercial effect. For Chrysler A-body, it was the 273 V8. A 2bbl device delivering 180bhp at 4200rpm and 260lb/ft at 1600rpm. According to the March 1964 issue of *Road Test*, it helped a four-speed 2.93 rear axled Valiant to 60mph in 9.9 seconds. The quarter-mile was crossed in 17.5 seconds at 79mph, allied with overall fuel economy of 15.8mpg. That LA small block V8 family would chug on to 2002!

Yes, the 273 was related to 1955's 318, but there were many differences made for A-body. Changes concerned the small block, and keeping A-body in balance for the power hike. Engineers wanted low weight, good performance/ economy blend, easy service access, and no special engine compartment changes to accommodate the V8. Thus 273, unlike 318 V8, had a single level intake manifold and lower air cleaner. The left exhaust manifold was a custom job to ease V8 fitment, and a shallow pocket was cast into the left hand side of the 273 V8. Versus 318 V8, precision casting was employed to minimize block material and make 273 lighter. It weighed 70lb under big brother 318. To lower production cost, 318's polyspherical combustion chambers and staggered semi-inclined valves gave way to 273's wedge-shaped chamber and inline upright valves.

The 273 conversion was no drop-in and forget job. The front torsion bars were still 90lb/in for firmness, but rear leaf springs were raised to 110lb/in. Slant-6 A-bodies rode on 6.50-13 tires, but V8's used heftier 7.00-13s, so heavier valved dampers were fitted. Chassis stiffeners curbed flexing, and larger radiator, plus upgraded TorqueFlite, were pressed into V8 service. The latter had the 225's ratios, but added a bigger torque converter, heavier clutches and discs, and higher gearbox oil capacity. Unfortunately, like Dart, they didn't improve the brakes versus I6 cars. One just had to hold on and pray for the 1965 disk brake option.

As a sign of increasingly higher performance times, Imperial was the only Chrysler Corp car one couldn't get a four-speed in! More than that, in the standard size 116in wheelbase Plymouth line, and 119in Dodge domain, that four-speed could be teamed with two Hemi 426s! The 'street' job made 365 horse handy for going to Sunday church ... or, there was the 425bhp, dual-quad job; with this hot tamale, one needed to visit church everyday to repent! Plymouth suggested the sportier 426 be used for Super Stock comp only. The valve timing, with lots of overlap, made it a real bear when getting groceries. You couldn't option this V8 in an A-body yet. Buyers also couldn't get Imperial's delicious five position reclining front seats, however, better stuff was on the way for '65 MY.

1965 was a styling carryover year, but under that skin, Valiant could win! The newly optioned 235 horse 10.5:1 CR Commando 273 V8 could be combined with four-speed, Sure-Grip lsd, and aforementioned disk brakes to create a budget C2 Vette 327 rival; not such a far-out scheme! Plymouth had fielded a successful Valiant two-door sedan-utilizing rally team in the 1965 and 1966 SCCA Manufacturers Rally Championships. Its star driver was Scott Harvey. Rallying helped development of Valiant V8 and Formula S Barracuda. As Harvey told *Sports Car Graphic* in December 1967, a six-cylinder Valiant with 120lb of lead in the nose had been trialed to simulate the V8's extra weight. In reality, the 273 V8 only weighed 50lb over the 225-6.

The Barracuda Formula S pack first started out on 1962's Valiant Red Snapper prototype. The Valiant V8 was even used to good effect at the 1963 Monte Carlo Rally. In great PR, a Valiant V8 came third in class. It even set the fastest time in one heat of a speed test conducted on the grand prix circuit. In such international events, Scott Harvey had Finland's Esko Keinenen as a team-mate. Said Finnish driver promoted Valiant in his country. Rally legend Timo Makinen also raced the Valiant in Finnish rallies.

Of greater import to American buyers was Chrysler Corp's 50,000 mile power train warranty. It helped dispel the corporation's recent reputation of poor reliability. The warranty went hand-in-hand with the public perception that Chrysler A-body was a solid, good value buy. Plus, spare a thought for what Chrysler's future corporate masters were doing in the mid '60s: in Mercedes' early postwar economic times it had a distributive link-up with Studebaker-Packard – by 1964, its Grosser 600 was a 126in wheelbased, 6.3-liter fuel injected V8 Goliath, weighing 5500lb. Fiat was regarded as the GM of Italy; the family cars it sold in North America, indirectly, via importing agents, were very small by US standards.

Try a rear engine 78in wheelbased Fiat 600D, or four-door 1100-B sedan on a 92in wheelbase. For la dolce vita, what about the Fiat Spyder two-door sports convertible? They were made in Italy in those days! However, most import buyers didn't look past the quality and reliability of VW Bug. For the same reasons, American economy buyers trusted Valiant. 1966 brought more noticeable visual change. A new split grille with fine pattern insert, new front fenders, and sedan rear fenders too. There was a beveled edge to the rear deck, and heftier rear bumpers. 1966 would prove to be the final year for the 106in wheelbase Valiant. 1967 witnessed a major restyle and re-engineering of A-body: a simplified range of fewer versions, which knew their target buyer;

Thanks to Esko Keinenen and Timo Makinen, the Plymouth Valiant V8 had a high profile in 1960s Finnish rallying. (Courtesy Kari Partanen)

Whether it was the 180bhp 273 or hi po 235 horse Commando unit, the Mopar LA V8 had the measure of Ford and GM's small blocks. (Courtesy Kari Partanen)

The V200 Signet implied buckets, but there was a V200 ragtop minus the sporty seats. Valiant could be custom trim optioned like any pony car. (Courtesy Kari Partanen)

A 1965 Plymouth Valiant V200 Signet V8. Valiant offered the cheapest V8 convertible on the planet! (Courtesy Kari Partanen)

wagons, hardtops and convertibles were gone.

Valiant continued in two- and four-door sedan form on a 108in wheelbase. Production continued at Belvedere, Hamtramck, Los Angeles, Delaware, Missouri, and Canada's Windsor, Ontario factories. Even economy Valiant was a dedicated follower of fashion. The boxy fendered, slab-sided look was very 1966. The vertical split grille was subdivided horizontally, and vertical taillights were segmented. The base 170in^3 six was upgraded from 101 to 115bhp through a carb-and-cam change. The 225's bigger 1965 spec camshaft and Carter BBS or Holley 1920 units with the 225's 43mm throttle bore were now employed. The 170in^3 six had used a 40mm throttle bore.

The majority of changes for the next few years centered on federally mandated auto safety and emissions. In 1967, that meant dual-chambered brake master cylinder, energy absorbing steering column/wheel crash pad, safety dashboard switches, shoulder belt mountings for driver and front passenger, four-way hazard flasher and directional indicators. For '68 MY Valiant lost the front fascia grille dividing bar, but got a fine crosshatch insert with segmented chrome surround. Model nameplate callouts were also moved from the rear to front fenders. There was the fed-required amber front and red rear sidemarkers. From January 1, 1968, front shoulder belts were mandatory.

On the excitement front, Valiant mirrored Dart's 318 option and had the hi po 340 as a special order V8. For '69 MY a one-piece full-width grille joined new taillights. The former Clean Air Package was renamed Clean Air System, with revisions netting better six-cylinder gas mileage. Brake adjusters were improved, so was the efficiency of the power steering pump. Chrysler Corp claimed to have improved its Sure-Grip lsd, and front head restraints were standard. Moving to 1970 saw the 170 replaced by the 198in^3 six, which shared the 225's block, and detail styling changes. The metal front grille was

A 1966 Valiant 225 two-door sedan, with three on the tree and zero options. Still featuring original paint and power train after 126,000 miles. (Courtesy Zach Thomas)

In 1932, Chrysler Corp spent $9 million on factory re-tooling as it transformed its inline four into an inline six. For 1960, the motor went from L-head to OHV layout in the new Valiant. (Courtesy Zach Thomas)

restyled into a black plastic unit. The center part of the grille was now flush with the forward edge of the hood. The remaining part of the grille was recessed. Valiant also enjoyed improved sound proofing in '70 MY.

However, Plymouth made some noise with the new Duster coupe. The Duster replaced Valiant's two-door sedan, and featured greatly curved side glass. It represented one half the curvature radius of the sedan's glass. Complete with muscular C-pillar, it looked like a mini muscle car; and that was the idea. As *Road Test* said in May 1975, Duster was dreamt up in the late '60s height of the muscle car era. Doubters of the era's influence should know that *Motor Trend*'s 1969 Car of the Year was the Plymouth Road Runner 383. A few years later it would have been a front drive I4 econobox but, for now, stripped-down, no-frills Mopar V8 muscle held court.

Motor Trend didn't stop at the 383 V8, no sir. In its December 1969 'A Date With Three Strippers' test, it lined up Road Runner 440 six pack four-speed against Chevelle SS 454 and Torino Cobra. The mighty Mopar had Air Grabber cold air induction, and far out 4.10 rear axle. Born for the quarter-mile, but this RR was more 'sleep, sleep' than 'beep, beep.' So, the legendary Norm Thatcher was drafted in to save Road Runner from Wile E Coyote! Thatcher found sticky valves and lifters. After fixing this, he richened the six pack's center carb and recurved the dizzy. The result was a 14.06 at 101.69mph. Other race tricks employed to lower that sacred E.T. were pumping up the rear Polyglas tires to 36psi, and slackening the fan belt. After all, America did discover the drag strip way before it found the Food Network.

The new Duster gave off this muscle car vibe. It didn't matter if it had the secretary special 198in^3 six, or top hi po 275 horse 340 V8. Those coupe looks modernized what was considered an increasingly staid Valiant. Duster was a showroom magnet that helped Valiant sales all along the line. Plymouth had a coupe exclusive in 1970 too.

The 1968 Valiant featured more V8 moxie. The 230bhp 318 V8 became a new regular option. The hi po 275 horse 340 V8 job was special order. (Marc Cranswick)

The quarter-mile figure at Orange County International's drag strip was none too shabby either. Here, the 108in wheelbase Duster out-duked its 111in wheelbase Swinger 340 sibling. It was a 14.22 at 99.88mph versus a 14.7 at 96.84mph, according to *Car Craft*'s May '70 report. At $2547 Duster 340 was considered superb value; its base price included what rivals had as options. On the negative side, the short 108in wheelbase made handling a touch squirrely. Bumpy roads could easily knock Duster offline.

Duster adopted a shark tooth grille in 1971, and was always an insurance premium inflation fighter. With the cost of sporty cars going up every which way, Duster stopped one's budget going loose. Enter '71's Duster Twister package: maximum sport on a tight budget, with an emphasis on appearance. Duster's stripes were complemented by Twister's decals, Rallye rims, dual racing mirrors, dual hoodscoops, flat black hood paint, strobe stripes, plain cloth and vinyl trim. There were four exterior hues to choose from, but the biggest Twister mill was the 318 V8. The 318 stopped one racing into debt, and according to Jim Brokaw of *Motor Trend*, had sufficient street moxie.

Motor Trend's 'The Invisible Cars' June 1971 report lined up a Satellite 318 against Torino Brougham 351 and Chevy Malibu SS 350. All three cars with low lead, compression drop V8s could still chirp tires in a midsize. Plymouth's smaller 2bbl V8 made 230bhp at 4400rpm and 320lb/ft at two grand, on an 8.8:1 CR. It delivered the best gas mileage on test, and nearly matched the bigger V8s for scat. 0-60mph times for Malibu, Satellite and Torino were 9.7, 9.9 and 9.6 seconds, respectively – and don't forget, Chrysler Corp were using said 318 in Indycar racing! Chrysler were trying a couple of new strategies at the time. Interior space efficiency was granted a higher priority relative to styling, and quality control was under review.

The corporation's quality had been slipping since the late '60s. Assembly line fit-and-finish, along with reliability complaints, caused action. Former Dodge General Manager Byron Nichols was put in charge of a special program to address such concerns. Across the street, AMC was getting a quality bad rap, and responded with its famous Buyer Protection Plan. By 1975, Chrysler Corp had its 'Clincher' 12 month unlimited mileage warranty, covering everything bar trim wear. Improved quality and better warranties helped sell cars in Detroit at this time, especially with economy car buyers. The less that goes wrong, the less it costs. For years, VW Bug majored on good warranty and few defects. Buyers wanted that peace of mind.

In the 1971 TV movie *Duel*, salesman David Mann's Plymouth Valiant sedan is pursued by a souped up diesel truck. The film starred Dennis Weaver, and was directed by Steven Spielberg. (Marc Cranswick, youtu.be/eLr2LRqTLwo)

This 1972 Duster 440 is Plum Crazy Purple. The Duster was conceived at the height of the muscle car era, making this hue very apt. Dusters rested on a 108in wheelbase. (Courtesy Frank Orlando)

The 440-4 ain't original. The big block V8 sports an 800 CFM Edelbrock carb, headers, and high rise Edelbrock intake manifold. Big blocks in compacts spelt 13 second quarters, or less! By 1971-72, Duster's Dodge sibling was Demon. (Courtesy Frank Orlando)

1971 saw a change in Chrysler Corp policy; interior space was given higher priority than styling. Sadly, on the A-body coupe this meant limiting front seat travel to maintain rear seat accommodation. This Duster features the very effective factory a/c. (Courtesy Frank Orlando)

Dragstrip-ready with Lakewood traction bars, this modified Big Block Duster resides in Fort Myers, Florida. The Duster and Demon are appreciating collector cars. (Courtesy Frank Orlando)

This 1973 Plymouth Duster 340 was imported to England from Orlando, Florida in May 2016. It used to have an H code 340, but is now fitted with an owner-built, bored and stroked 418in^3 V8, making 475 horse and 500lb/ft. A Scat 4in stroker crank and H-beam con rods are present, using ARP bolts. (Courtesy Steve Waters)

1972-76 – it's tough being a compact

There was a policy change concerning high-performance, too. As the big blocks bowed out at the close of '71 MY, the once high-strung 340 V8 moved into the mainstream, sporty, all-purpose V8. It kind of worked; come 1972, one turned to the 340 V8, even in Charger class. In Jim Brokaw's 'Sayonara Supercar' June '72 *Motor Trend* article, he said 340 V8 had become a reformed citizen. No longer bogging off the line, but possessing a smooth mid-range that could match Glen Campbell on High-Fidelity 8-track. The top end wasn't what it once was, but, then again, little of anything was anymore. Describing 'the old days' (as recently as 1968, would you believe?) Brokaw recalled "... Engine compartments were stuffed tighter than a Laguna Beach bikini ..."

Now things were returning to the early '50s, when kids jazzed up tired old iron with trinkets and baubels. In the early '70s, the term 'paint & tape specials' started to get airplay. Not that the majority really cared anymore, but Chrysler Corp still cared: describing Mopar's continued magic, Jim Brokaw said "These are not cars you like, these are cars you love," before countering: "It's too bad that people don't seem to love their cars anymore ..." He was speaking of Mopars in particular, and high-performance cars in general.

Proof of the public's indifference was its response to the Duster 340 in 1973-74. Chrysler Corp had reconfigured the 360 V8 as a performance replacement for the 340. On the new Duster 360, one could have a Hurst shifted four-speed and 3.55 rear gears, as before. Weight was around 150lb heavier compared to 1971 Duster 340, mainly from fed impact bumpers and mandatory steel door beams. The latest Duster was a little tardier, but that wasn't the main reason folks were passing on this ride, or

As part of the 418's build, Icon 744 forged flat top pistons, and ported 1971 J heads with 2.02in inlet valves and 1.60in exhaust valves were specified. A Comp Cams XE 274 H 242 duration cam with 0.492in lift keeps things lively. Felpro 521 SD high-performance head gaskets help maintain the 10.75:1 CR. (Courtesy Steve Waters)

Challengers and Z/28s, etc: the fuel crisis, with rising gas prices and shortages, inflation, car insurance, and increased interest in practical matters, slowed sporty car sales.

Just as well Valiant still majored as a solid American economy car. In 1971, the compact Plymouth sold 'only' 256,930 units on calendar year sales. 1972's tally was a record 330,373 cars. While dust gathered on full-size lead sleds, the A-body's value for money and reliability, especially in Slant-6 form, made it a continued strong seller – ask 381,000 mostly pleased buyers in 1973. In 1974, Valiant racked up 477,000 sales as VW Bug floundered. Who was the economy car king now?! In tough times, Valiant was your friend. Even so, annual revisions

The factory hoodscoops have been made functional, delivering cold air to an Edelbrock Victor 340 high rise intake manifold, and Holley 750 CFM vacuum actuated secondaries 4bbl carb. Mopar Performance cast aluminum valve covers dress up the small block. (Courtesy Steve Waters)

weren't exactly enthusiast-aimed. 1971 had seen EGR and a charcoal filter arrive to fight pollution. 1971's new Valiant Scamp was an 111in wheelbased, badge-engineered equivalent to Dodge's Swinger two-door hardtop. It carried Valiant's front fascia, but Dart's 1970 dual oblong horizontal taillights also rode shotgun. By the early '70s, Valiant accounted for over 40 per cent of Plymouth sales.

1972 saw a revised grille and taillights, plus surface mount sidemarkers replacing previous flush mount types. 1973, a bumper sales year for the industry, saw fed-mandated front 5mph impact bumpers, utilizing a telescopic mechanism and rubber boot cover. There were no vent wings on Scamp anymore, and 1973 was the last year for Valiant's 108in wheelbase, bar A-body coupe. By 1974, economies of scale dictated a badge-engineered version of Dart, with that car's 111in wheelbase. The revisions saw stouter C-pillars, new rear fender sheet metal, albeit with Valiant retained '73 front grille and styling. 1974 implied chrome front bumper guards, and 5mph rear impact bumpers, plus a four-door Scamp

This Duster's interior has new seat covers, carpet, and roof lining. Center console and dashboard have been refurbished, with auxiliary gauges and tach added. Safety rests with factory front disks, aided by a new power booster and master cylinder. (Courtesy Steve Waters)

Further engine hardware include Clevitte H-series racing rod/full groove main bearings, Elgin high-volume oil pump, and 8qt baffled Moroso oil pan. Power goes to an A727 TorqueFlite with reverse manual valve body, Turbo Action 3500 stall converter, and Mopar 8.75in differential with 3.23 ratio Sure-Grip lsd. (Courtesy Steve Waters)

To boost rigidity, this Duster has subframe connectors. A propshaft safety hoop, rear tubs, and Competition Engineering adjustable drag shocks join rear springs that have been moved inward. Mopar Rallye rims are sized 14x6.5in front with BF Goodrich 235/60-14s, and 15x8.5in rearwards with BF Goodrich 275/60-15s. A 12 gallon Ali fuel cell resides in the trunk. (Courtesy Steve Waters)

package. The pack had an eye to the downsized luxo many buyers wanted, moving from intermediate to compact to save $$$s.

Scamp's spec encompassed 318 V8/TorqueFlite, power steering, power front disks, chrome trim, vinyl roofs, a/c, deluxe carpet, split vinyl bench, radio, rear window defroster blower, and 120mph speedo; this last item was special order. Los Angeles Valiant production had been suspended. Continuing the luxury theme, Valiant Brougham teamed up with aforementioned Dart Special Edition. Brougham offered a choice of two or four-door sedan, standard chrome pack, vinyl roof, deep-cut pile carpets, velour cloth trim, interior door padding, color-keyed rims or faux wires, and special paint/trim combos. As expected, power steering, power front disks, a/c, cruise control, electric rear window defogger and AM/FM radio were all standard – no need to visit Caddy, Lincoln dealers or even option an Imperial!

Come 1975 and Valiant got another grille restyle, plus California and some high-altitude areas got catalytic converter equipped Valiants. Radial tires were standard equipment that year, and cost saving measures were afoot. The 'Fuel Pacer' dash light signified a vacuum gauge, to encourage a light foot. The overdrive version of the Chrysler four-speed (A833 OD) also spelt thrift. It was the first time since 1965 that a six could be teamed with four-on-the-floor. In keeping with the times, saving money and federal safety laws were high priorities. Spark plugs were supposed to last 50,000 miles, and there was that Clincher warranty. For '76, turn signal lenses became amber, not the former clear lens relying on an amber bulb. Finally, the hazardous under-dash park brake handle had been replaced by a foot park brake.

See you in Saint Louis

At this time, A-body assembly moved to Saint Louis, as Hamtramck was set up for the new F-body. However, Chrysler Corp stayed faithful to law enforcement agencies with its usual A38 Police Package. There were different engine choices, depending on whether the purpose was patrol or pursuit. With A-body, a 225 1bbl six was under the code E24 (California) or E25 (49 state). Midway brought the E44's two barrel 318. Hot pursuit implied the E58 four barrel 360 V8. California's package had

A 1974 Valiant sedan. By now, Valiant was twinned with Dart on the same 111in wheelbase. However, cosmetic differences continued. (Courtesy David Zatz)

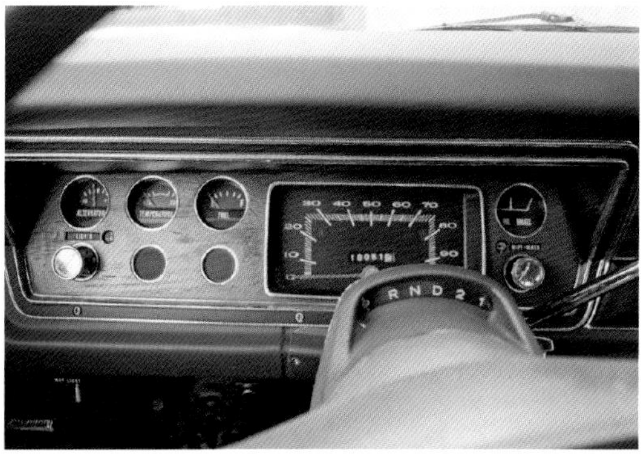

In the wake of the gas crunch, owners were studying the optional Fuel Pacer vacuum gauge dash light with great interest. The 225 Slant-6 and 318 V8 were thrifty for their size. (Courtesy David Zatz)

The smooth luxury of TorqueFlite auto and factory a/c. A 120mph speedo was a special order option. However, buyers were more into plush than performance by 1974. (Courtesy David Zatz)

At 202in long, was this the new full-size for the '80s? 1974 was the first year for federal 5mph impact bumpers front and rear. Handy for parking by ear! (Courtesy David Zatz)

a single exhaust system and cat, for 175 horses; however, 49 state flyers possessed 220 net ponies and no cat.

The Lone Highway Ranger's steed did mid-16s at 85mph, as long as he wasn't riding through the Golden State. The cops had qualms concerning A-body. Although visibility and handling were tops, the unibody lacked strength. Police A-bodies included A38's handling package of front and rear swaybars. They also borrowed strengthening mods from other A38 packs; namely, more body welds, and reinforced crossmember. Even so, Valiant's K-frame could crack under severe cop usage, so law enforcement stayed loyal to midsize and full-size during this era.

Ode to the A-body

On the question of performance, Chrysler Corp, and all other car makers, had been laid low by high-minded Washington bureaucrats. Detroit and others had largely been getting by, and even had a response to the compression ratio drop of 1971-72, but the heat was on (or off, depending on whether you liked fast cars). The 1970 Muskie Smog Bill had resulted in the higher standards Washington wished to see from Detroit. In April 1973, the EPA even investigated how well the domestics were doing with their homework. GM was almost done on paper, but Walter P was on 0 per cent concerning meeting proposed 1975 emissions requirements. Indeed, due to troubled times, Chrysler Corp had only been allocating one third the revenue share to R&D compared to GM and Ford.

Washington's attitude was 'here are emissions standards, as automakers it's your baby.' The trouble was, the legislators didn't understand that there was no effective, long-term approach applicable to affordable cars to deliver the drastic cuts in pollution Washington wanted. The company that made the first motorcar, Daimler-Benz, sold its 280S with 160 horses in Europe; in America, it was 120bhp. In the short run, manufacturers could tune their cars to pass, but only with much poorer power, awful gas mileage, and cantankerous driveability. Still, Washington's conscience was clear, but consumers got the burden. Engineers spent the next 15-20 years trying to get back to 1967.

In these trying times, the public still placed its faith in Chrysler A-body. So much so that in May 1975, *Road Test* tried to find out why. The journal got hold of a Dart SE sedan and plain Jane Duster, running them through the usual battery of objective tests. After doing so, *Road Test* was at a loss to explain A-body's great owner loyalty. Acceleration, handling, build quality, and fuel economy were par for the course; nothing exceptional, in the magazine's opinion. The mind was cast back to the Plymouth Fury that raced imported sportscars, and 300Cs that handled with aplomb. What happened?

Emission controls had dampened the enthusiasm of the 225 Slant-6. Impact bumpers added weight and length, which upset handling balance. The lithe 1963 Darts and Valiants had been under siege from legislators. *Road Test* questioned Detroit's preparedness for a future where fuel economy and quality control would be key. However, the reality was that no car in 1975 behaved like it used to. In the brave new world, A-body was the consumer's Plymouth Rock of stability. It was a known quantity in a world of unknown.

Doug 'Dr Dodge' Dutra noted the fragility of the aluminum 225 block's casting, saying regular coolant changes with anti-freeze were necessary to ward off electrolytic corrosion. Dr Dodge also said the best thing Chrysler did was to give the aluminum crankcase dimensions to regular 225s. It made the Slant-6 unbreakable! Chrysler's LA V8 was also still a prince among small blocks. *Road Test* concluded that people like A-bodies, just like they admire Mustangs, Camaros, and Caddies. It also had a theory that the average Duster owner was a secretary in Fresno. Well, millions of A-body owners can't be wrong; try 2,428,177 Valiant customers between 1967 and 1976: they heard there was a new model afoot, and they were turned on to a possible trade-in!

Chrysler's small cars did well in the fuel crisis. Import buyers went for Colt's quality, and domestic buyers chose around 477,000 1974 Valiants – this was one of them! (Courtesy David Zatz)

5

From F-body to Hellcat & Demon! – The Return of High-Performance

Aspen & Volaré – the classy F-bodies

The new Dodge Aspen and Plymouth Volaré had to succeed – given Chrysler's domestic and international problems, they just had to. Walter P was unlucky in having revised his full-size line just when buyers were downsizing to inters, or even smaller. The gas crunch caught many out, so, as ever, Dodge trucks and the A-body's replacement would have to suffice. In these changing times what used to be a senior compact was now a midsize or more. So Chrysler's hopes and marketing seemed spot on. The new F-body was billed as "The standard cars of the future."

The cars needed a new name. Dodge buyers were typically over 50. Indeed, demographics showed the domestic buyer profile from 1975 to favor older drivers – to some extent, traditionalists, who preferred a rear drive I6 or V8 sedan/wagon to shoehorning themselves into a pricey four-cylinder econobox. Product Planner Burton Bouwkamp, and Dodge Car Division Manager/Chrysler Corp Ad man RM 'Ham' Schirmer, wanted buyers to know that Aspen and Volaré were really new. So even the family code designation changed to F-body, sadly breaking Chrysler's logical small, medium, and large respective A-body, B-body, and C-body tags.

The strategy worked. Not too many 1976 Darts and Valiants went out the showroom door. A-body didn't see out the 1976 model year; buyers knew something new was on the horizon. They held out for the mid '76 MY intro of F-body to the 'Unbelievable' but true tune of 189,900 Aspens and 255,008 Volarés. What's in a name? Plenty, it would seem. After a lifetime of Dart and Valiant monikers, the new nameplates were in keeping with the upscale, continental, and lifestyle times. In Spanish, Volaré means 'I will fly;' in Italian, 'to fly.' The word's accent on the letter 'e' was kept. Ads played on the Spanish spelling, announcing Volaré had 'an accent' on quality – and that's not all it had.

Chrysler Corp didn't get Dean Martin to front Plymouth Volaré's marketing campaign. Instead it got latin lounge legend Sergio Franchi. Comedian Alan King once observed that Ed Sullivan had such a hard time pronouncing his name, he nearly experienced a self-induced infarction. After several failed attempts, he gave up and just said: "Folks, let's hear it for *Ave Maria*!" On radio and TV Franchi sang altered lyrics to the hit song. However, no mention was made of rich Corinthian leather. No need to, when Dodge Aspen's campaign was literally beyond belief. Aspen was promoted on TV and in print ads by British actor Rex Harrison.

The Dodge Aspen ad campaign was built around the slogan 'Unbelievable.' Harrison reprised his persona from the 1964 movie *My Fair Lady*, co-starring Audrey Hepburn. In a recreation of the movie's Ascot Gavotte scene, Rex Harrison performed the 'Unbelievable' song, in his inimitable talk/sing fashion; the actor didn't have Sergio Franchi's natural singing voice. By Jove, I do believe they've got it! Dart's transformation into the upscale, lavish Aspen was quite unbelievable, but true.

The Dodge Aspen's name arose from Aspen's Aspen-Vail project codename. Marketing considered it a pleasant, natural name, which wasn't solely associated with skiing, but was evocative of the great outdoors. Even so, Chrysler Corp did formally sponsor the K2 Skis demonstration team, as the '1976 Dodge Aspen K2 Freestyle Team.' In addition, Chrysler Corp was involved with the opening of World Pro Skiing's 7th annual season: the first round was held in Aspen, Colorado, as part of the Dodge Aspen Cup. So the sponsorship worked hand in ski! The classiness continued with F-body's substantial revisions to bodywork and suspension.

Re-engineered for refinement

The public desired downsized luxury, and Chrysler Corp made Aspen and Volaré very attractive propositions. The restyled bodies boasted a larger greenhouse, lower belt line, higher seating, lower steering wheel, and no rain gutters. Bodystyles were two-door coupe, four-door sedan and, for the first time since 1966, four-door wagon. The last variant was designed to challenge the domestic dominance of AMC's Sportabout compact wagon. Dart's popular two-door hardtop wasn't directly replaced. F-body coupe had 25 per cent more glass, and sedan, 33 per cent more greenhouse, than their A-body predecessors. Rear windows were more flush mounted, as part of wind tunnel assisted aerodynamic work carried out on F-body.

Such aero studies saw smoother sheet metal edges, better crosswind stability, and reduced wind noise, as part of a 10 per cent reduction in drag coefficient. It was all very timely for the 1978 commencement of CAFE fuel mileage targets. Computer Aided Design (CAD) also helped F-body become a lighter, but more rigid structure than A-body. Clear plastic stress modeling led to a body with fewer stampings, fewer welds for more strength, lighter weight, and improved panel fit – assembly line permitting. In keeping with automakers globally, Chrysler Corp looked at reducing weight to improve economy in the wake of the first gas crunch, so F-body had thinner glass, lighter side intrusion beams, along with the use of high-strength light alloys for brackets and general body beefing up.

On the inside, engineers improved fresh-air ventilation ducting hardware. They also added more sound deadening material, thicker carpeting, improved body door seals, and insulation. For an upscale ambiance an upper door chrome pack, covered windowsill line, and exterior reversing mirrors. There was an in-house realization that some Chrysler B and C-body sales would be poached by the new F-body. There were great design and planning similarities between A-body's F-body transition and AMC's Hornet/Concord metamorphosis.

In TV and print ads, actor Rex Harrison fronted the Aspen's launch campaign, with the 'Unbelievable' tagline. It all invoked the spirit of the popular *My Fair Lady* movie he starred in with Audrey Hepburn. (Courtesy Dodge, dodge.com)

F-body engineers redid Chrysler Corp's traditional suspension, so the new car would have lighter hardware and more compact layout, which also delivered a big car ride; conflicting goals that were mostly harmonized by Aspen and Volaré.

To use an umbrella phrase to describe F-body's new set up, an 'Isolated Transverse Torsion Bar System' was present. No longer longitudinal, Chrysler's famous torsion bars were now parallel to the K-frame. There was also a lot of

This 1977 Plymouth Volaré Premier was purchased from Gator Chrysler Plymouth in Melbourne, Florida. It is now a little north of 53,000 odometer miles. (Courtesy Gene Yetter)

rubber used throughout the front and rear suspension, for better refinement. The now transverse torsion bars had an anchor near the suspension's control arm, on the opposite side. The L-shaped torsion bars ran parallel and ahead of the K-subframe. This crossmember was attached to the unibody via four rubber mounts. Ride height could still be adjusted for suspension wear via the torsion bar system. Removing the wheel housing plates permitted camber and caster adjustment.

Still with a leaf-sprung live axle rearwards, but an Iso-Clamp rubber doughnut between spring perch and body aided front to rear cushioning. The NVH reducing Iso-Clamp design, was applied to all rear drive Chrysler Corp cars from 1976. Even the steering column was rubber isolated. All chassis upgrades applied to Aspen and Volaré's three trim levels; both came in base or 'Custom' trim, with a top trim level of 'SE' for Aspen, or 'Premier' for Volaré.

Although the coupe bodyshape was present from the F-body's introduction, it took some time for the Duster to formally arrive. A top seller in A-body days, the Duster wasn't considered 'in keeping' with F-body's upscale mission. However, seeking more sales in competitive times, the famous Duster name was dusted off for '78 MY. The Volaré Duster was denoted RPOA42. Some felt the Duster style to be incongruous with the classy F-body. It wasn't as popular as before, but it did sell well.

There was certainly a switch to lavish editions. F-body was Chrysler Corp's top seller in 1977, and Canada's best selling car that same year. Aspen sales were split 109,674/66,844/89,494 for its base, Custom, and SE (Special Edition) respectively. Volaré ventured forth with 139,865/75,360/112,514 for respective base, Custom, and Premier.

So, buyers really dug the pricey SE and Premiers. Dealers loved the higher mark-up; Volaré Custom Coupe was

After the early '70s, buyers were looking for downsized luxury, and finding it in the factory velour interior of this Volaré. (Courtesy Gene Yetter)

The Super Six two-barrel 225 Slant-6 was a $40 option. Popular and desirable, the Super Six package brought useful upgrades, beyond just the 2bbl Carter BBD. (Courtesy Gene Yetter)

$3671, but Premier was $4515. Those stripper, bench seat, and glass pak days were gone! Size-wise, one's coupe was 198.8in long, just like a GM F-body, 73.3in wide, and 53.3in tall. Trunk capacity was much higher than the GM coupe, at a huge 16.6ft³.

Six-cylinder cars had an 18 gallon tank, and 7¼in differential. With V8s it was 19½ gallons and an 8¼in differential. Towing packs also procured the bigger diff. Rear axle ratios ranged from a sleepy 2.45, through 2.76 and 2.94, with Sure-Gripped 3.23 as the performance choice. Lightest was a Slant-6 coupe at 3200lb, with 110lb more for a V8. Four-door sedans added 75lb over the two-door, with wagons 200lb heftier than four-door. With an 1100lb load capacity, F-body was only 100lb shy of a large Chrysler wagon. In downsized time, F-body looked like America's new full-size. With the convenience of a liftgate, wagon

So popular were smaller luxury cars that Volaré Premier's 112,514 1977 sales exceeded those of Volaré Custom handsomely. (Courtesy Gene Yetter)

A cloth Landau top made this Volaré Premier anything but an economy car. With 327,739 sales, 1977 was Volaré's best year. (Courtesy Gene Yetter)

A 1977 Plymouth Volaré Suburban wagon: F-body four-door sedans and wagons rode on a 112.7in wheelbase. Chrysler Corp billed the F-body as representing "The standard cars of the future." (Courtesy Bud Ward)

offered 71.9ft^3 of load space. With the rear seats folded down, a handy 74in x 43.2in presented itself between the wheelhouses.

Cargo side cover panels were made of durable one piece injection molded polypropylene. Standard on Aspen SE and Volaré Premier wagons were lockable stowage compartments, worth an extra 1.5ft^3. Both upscale wagons came with the usual faux exterior woodgrain paneled decoration. On Aspen SE, it was woodgrain as framed in simulated painted metal trim; on Premier, stainless steel frames had a matte black accent. *Road & Track* sampled a Volaré 318 wagon and recorded 0-60mph in 14.6 seconds. The zombie shift carry-all managed 60mph to zero in 198ft, and a skidpad figure of 0.668g. All fine for a 3960lb family hauler.

In 1977-78, The F-body's base wagon motor was the 2bbl 225 Super Six. This implied the 2.76 to 2.94 rear axle change that Super Six brought. However, this wagon sports the 145 horse 318 2bbl V8. (Courtesy Bud Ward)

F-body challenges – reliability and economy

Motor Trend was pretty impressed with F-body. Introduced in the fall of 1975, the magazine named Aspen and Volaré its joint Car of the Year, 1976. However, not everyone was impressed. Apart from the usual assembly line snafus accompanying any new model, F-body had paint problems and rust issues. Such '76 MY glitches caused Chrysler Corp to ditch its usual asphalt based paint and protection method. In came a new seven stage autophoretic coating system. It was a first for domestic automakers. Metal panels were spray cleaned, dipped clean, rinsed several times, and recirculated in fresh water, plus de-ionized water. Panels then got the autophoretic coating as a deep chemical rinse, followed by a reaction rinse. Panels were oven dried for 5-10 minutes at 220F, then cured for 5 minutes at 275F.

The corrosion issues were mainly

with the front fenders. Under a recall, many 1976 and 1977 F-bodies received new galvanized steel front fenders. Plastic polyethylene inner fender liners were included in the recall. It should be noted that, in 1980, Honda had a similar problem concerning rusted-out front fenders on 1975-78 Civics and Accords. 700,000 North American Hondas were in the recall. As expected, cars most affected were in colder climes. F-body recalls also concerned the ignition and fuel delivery systems. Rubber parts concerning the latter broke down too rapidly. Some of the F-body's problems were shared by all cars of its era, partly induced by legislation that focused on emissions, weight saving, and cost-cutting targets. It placed a burden on engineers and component suppliers. They just didn't make 'em like they used to!

The abovementioned problems tarnished F-body's reputation. Aspens and Volarés did improve, however solid A-body was a harder act to follow than Vega to Citation. F-body still sold in big numbers, with 1977 seeing 266,012 Aspens and 327,739 Volarés find homes, so they were popular on the street. Twins under the skin; sedan and wagon shared a 112.7in wheelbase, but Aspen had horizontal grille dividers. Volaré was well known for its egg crate affair. By 1978,

The Volaré Suburban name referred to a base-level wagon. However, options fitted include rear window defogger, AM/FM radio, and 2.94 rear axle with Sure-Grip lsd. Between 1977 and 2015, this 38,000 mile Volaré Suburban spent its life in New York. (Courtesy Bud Ward)

Aspen and Volaré sales had dropped to 166,419 and 217,795 respectively.

Sales declines were partly due to quality problems, but also the hot-selling new GM downsized '78 midsizers. Fairmont didn't help either. Also, F-body wasn't exactly a new design. Many families were downsizing into the 'Rabbit' four-cylinder, econobox market. F-body building continued at the usual locations of Hamtramck, Michigan, Bogota Columbia and Mexico City. A popular F-body addition in gas saving times was the Super Six option. In 1977 and 1978, the trusty 225 Slant-6 was available in 2bbl form for $40. The upgrade was engineered by Pete Hagenbuch and staff. The 1bbl Holley 1945 carb was replaced by a 2bbl Carter BBD, similar to the 318's unit. Early on, an aluminum intake manifold was specified, however poor quality casting saw a switch to a cast iron unit.

Unusually optioned in B-5 French Racing Blue, a shade that was one of the 'High Impact' colors usually reserved for F-body coupe. (Courtesy Bud Ward)

A 1979 Plymouth Volaré coupe. The Duster coupe body shape was considered not quite in keeping with F-body's upscale nature. However, it was still a strong seller. (Courtesy Ronald Neuendorf)

Available from the start in base, Custom and Premier trim levels, the formal Volaré Duster package was a late arrival. (Courtesy Ronald Neuendorf)

With the transverse torsion bar front, and Iso-Clamp rear suspension, F-body promised big car ride comfort. (Courtesy Ronald Neuendorf)

By 1979, Volaré was down to 178,819 sales, and Chrysler Corp was anxious for economical K-car to take up volume sales, post energy crisis II. (Courtesy Ronald Neuendorf)

The Carter carb had dual simultaneous opening 1.44in throttle bores. On the exhaust side was a 2¼in system, and mini three-way cat for California. Upgrades also encompassed a higher performance torque converter. The lighter 6lb battery was more compact, had better cell connectors, and vibration protection. A new terminal wiring system, and also new double contact starter relay brought improved cold starting. The two-barrel 225 was a modest 10bhp stronger, but better throttle response, and the associated upgrades, made it a desirable option. Buyers looked out for that "Super Six" air cleaner decal callout.

For so long, six-cylinder motors were ignored in Detroit; they were just an economical stepping stone to the base V8. Now, the fuel crisis and inflation challenges saw engineers redo their sixes for power and economy, as buyers searched for smaller engines. F-body's initial line-up involved a 1bbl 225 with 100bhp, 318 2bbl on 145bhp and Carter two-barrel 360 boasting 170 horse; torque ratings for these were 170, 250, and 280lb/ft, respectively. For Slant-6, base gearbox was the A230 three-speed column shift. The 318 came standard with three-by-the-knee. Either motor could have the overdrive A833 four-speed.

Automatic choices varied with torque at hand, between A904, A998 and A999 three-speed TorqueFlite boxes. There was no point having an HD gearbox handling a modestly powerful engine. In these crucial weight-saving times, rear spring hangers for 1978 were one half-pound lighter. 1978 also brought ELB (Electronic Lean Burn) to Chrysler Corp V8s. The first use Detroit had for computers underhood was ignition timing. For optimal economy, driveability, and power, a microprocessor could do a better job than points ignition. The term 'black box' entered the late '70s auto lexicon. In 1979, ELB was renamed ESC (Electronic Spark Control). Curiously, it was also the name GM chose for their system.

Early versions of ELB had two electronic pickups in the distributor – later examples, only one. The ignition black box was mounted on the air cleaner. In these smogger times, a Regular Fuel package was available for 318/auto combos. It used an air pump, but no cat in 1976-77. This 49 state pack allowed the driver to skip pricey, and hard to find, unleaded. The latter quality was a product of a lack of oil refinery capacity. The factory CB radio options were also a product of their times. Made by former defense contractor Huntsville Electronics Division, their AM/FM 40 channel CB Transceiver allowed one to join the trucker crowd, with the right handle. That's a big 10-4, Mopar One!

Watching out for Kojaks with Kodaks and bears in the air wasn't a big issue in Europe, although the Swiss were clean air fiends. Honoring F-body's upscale nature, Swiss car maker Monteverdi made the Monteverdi Sierra in four-door sedan and two-door cabriolet forms. Better known for dropping 440 and Hemi V8 mills into an exotic Euro chassis, Monteverdi's Sierra was a rebodied and re-engineered Chrysler Corp F-body. Styling was worthy of Pininfarina, with that small waisted Gina Lollobrigida look, and a fresh interior. Monteverdi gave Sierra a standard 318 V8, with the 360 optional. The Sierra was very expensive, exclusive, and had a reworked version of F-body's transverse torsion bar front suspension.

The Monteverdi Sierra sedan of 1977 was a rebodied Chrysler F-body, with optional 360 V8. A Dodge Diplomat-based cabriolet complemented the sedan. (Courtesy monteverdi.ch/en_museum.html)

1976 was the first year for the Dodge Aspen R/T and Plymouth Volaré Road Runner performance specials. A 1977 example of the former is displayed. The R/T (Road & Track), evoked nostalgia of yesteryear's hot Dodges. (Courtesy Eric Bremer)

Aspen R/T & Volaré Road Runner – last of the muscle cars?

F-body was less than 2in longer than A-body, but at 198.8in for coupe and 201.2in for sedan and wagon, they were increasingly large cars. The EPA offered consoling gas mileage numbers of 18/30mpg for wagon, and curiously lower 18/27mpg for the lighter sedan and coupe. Thank heavens for the 225 Slant-6. Aspens and Volarés had VINs commencing with 'N' and 'H,' respectively. By 1978, there was one trim level, with Custom, SE and Premier as option packs. In an auto scene where F-body appealed to traditional buyers, and many were attracted to four-cylinder import econoboxes, could there still be a pony/muscle enthusiast V8 rear driver?

Chrysler Corp thought there could be two, and called 'em Aspen R/T and Volaré Road Runner. The respective Dodge and Plymouth suffixes had an historical reputation that preceded F-body. Could hot Aspens and Volarés do justice to such monikers? Like any F-body, the Dodge and Plymouth versions were twins under that fancy trim. Their respective 1976 surcharges were $196 to make R/T, and $205 for Road Runner. The upgrade only applied to F-body coupes. A continuation of the hot Dart Sports and Valiant Dusters, if you will. The extra money procured a 318 V8, HD suspension with rear swaybar, and E70-14s for the familiar Rallye rims. Good value for the times. However, most Aspen R/T and Volaré Road Runner buyers also specified a Décor Pack: this delivered a front air dam and rear spoiler, lower-third body profile graphics, dual remote mirrors, headlamp trim, special horn and those iconic rear side window louvers. That last addition brought style, but did no favors for rear blind spot visibility.

In February 1976, *Motor Trend* timed an Aspen R/T 360 automatic coupe at 8.6 seconds for 0-60mph, and 17.4 seconds in the quarter-mile at 86mph. (Courtesy Eric Bremer)

The 318 implied four-speed fun, but the optional 360's honest 170 horses delivered performance equal to the best cars of the day, import and domestic. *Motor Trend*'s sampled Aspen R/T of February 1976, turned in 0-60mph in 8.6 seconds, quarter-mile in 17.4 seconds at 86.1mph, 115mph top speed and 15.2mpg overall; all with 2bbl carb and mandatory auto. *Motor Trend* titled its report: 'Aspen R/T – Rapid transit returns: an all-new performance car for the age of economy.' They were different times, where overall ability, not just acceleration, counted. At the time, even AMC PR folks said their new '77 AMX wasn't in the old muscle car idiom, but a balanced, sporty, personal car.

Back in November '68, *Car Life*'s '69 'Cuda 340 took on a super German and triumphed. These days, ten grand only got one a BMW 530i. *Car and Driver*'s equivalent stats were 8.7 seconds/17 seconds at 81.9mph/115mph/18mpg. The same magazine's April '77 figures for the 200bhp W72 400 powered Pontiac Trans Am were 9.3 seconds/16.9 seconds at 82mph/110mph/13.75 mpg. The R/T and Road Runner twins cost less than the Pontiac, too. Both BMW and Pontiac cars tested had four-speeds. A normal L82 350 motored 'Vette of the day was a mid-16s car, with zero to sixty in just under 9 seconds – once again, a pricier quasi-rival. That's latin for 'close, but no cigar!'

Chrysler Corp had packages covering the full spectrum, from paint to performance. In Volaré Duster's realm was the 1978 Sunrise package, a coupe with special stripes and aforementioned rear window louvers. Then there were coupes called 'Fun Runner,' 'Sun Runner,' and even 'Front Runner;' all cosmetic adjuncts to Volaré Road Runner. They had variations in decals, and used the familiar spoilers, fender flares, and window louvers, sometimes taking advantage of the T-Top option, available on coupes from 1977. T-Tops were a contemporary fad, promising sedan strength and convertible flair. However, man and gal can't live on paint & tape alone! Dealers used cosmetic packs to steer custom their way, but tick the A67 Super Coupe order blank and a real Mopar superhero was coming your way.

In its February '76 issue, *Motor Trend* harked back to the glorious past with its Aspen R/T headline: 'Rapid Transit Returns!' (Courtesy Eric Bremer)

The A67 Super Coupe, and add-on A43 'Street Kit Car,' were 1978-only performance and visual enhancement packs. They provided the ultimate in North American street scat. The ad copy was correct: "The Super Coupe Is Alive & Living In Aspen." Well ... make that Volaré Valley too, because Road Runner could have the package as well. It must have been worrying to John Denver and other environmentalists that a toxic F-body had taken up residence in their zip code! But it was all fair and legal. Super Coupe wasn't above the law; at A67's heart was the 360 4bbl V8 E58 police motor. Use of the bigger A727 TorqueFlite meant no duals. Just a single silencer and cat, ma'am, but all the other good stuff was there for cop roll call, so be careful out there ...

All Aspen R/Ts and Volaré Road Runners were V8 powered. However, from the factory, four-on-the-floor stopped at the 145 horse two-barrel 318. (Courtesy Eric Bremer)

Original specs for this 20,828 mile '77 Aspen R/T involved 2bbl 318, four-speed and 3.23 rear axle. It has since been given a 4bbl carb. (Courtesy Eric Bremer)

Semi-gloss black valve covers, Nimonic alloy exhaust valve heads, stiffer valve springs and high-strength rocker arms belied what must have been an underquoted 175 horse at 4600rpm. Silichrome-1 high-temperature steel intake valves lived in custom heads. There were special valve stem seals, double row timing chain, and optional oil cooler. An 800 CFM Thermo Quad harked back to the '71 Cuda 340s, and complemented a mild cam. The better cooled heads, and orange silicone rubber head gaskets, were amongst the performance and durability boosting goodies.

Although a cop motor, it was available to the public. Not that many were interested in such constrained times. The special 360 continued into '79 MY, getting stronger at 195 horses. Tightening emissions and CAFE, plus recession, ended it at that point. However, in 1978, Super Coupe added to the serious fun with an even heavier duty A43 pack: extra HD suspension, Aramid fiber GR60-15 Goodyear GT radials on negative offset 15x8in rims, with chrome lug nuts and chrome trim rings. The big wheels made the flared fenders truly functional. These flares, plus headlamp doors, grille, bumpers, windshield surround, door handles, gas filler cap, hood, roof, rear window louvers, and front and rear spoilers, had the popular matte black paint effect.

Tri-color stripes of orange, yellow and red set off the 531 Sable Tan Sunfire Metallic and 494 Crimson Sunfire Metallic Volaré Super Coupes. Naturally both cars carried 'Super Coupe' decal callouts. Windshield locks and hood pins recalled racer days of yore. The cars were instant collectibles, since it was originally planned to limit production to 1000 cars. Don't let that ELB air cleaner fool you – independent tests proved Super Coupe to be a high-16s car, with 0-60mph in 8 seconds flat. For those seeking more, the A43 'Street Kit Car' pack would provide.

Unlike A67, the A43 pack required power steering and power assist for F-body's 11in front disks/10in rear drums as mandatory options. The motor was a warmed up E58 360 four-barrel-of-fun V8, but still 49 state legal. Cast iron headers, shot-peened conrods, and high-volume oil pump took one back to 1968! Under the supervision of Michigan's police department, an A43 coupe managed 148mph. It took 12 miles to reach this terminal velocity, and the mid-15s quarter-mile also suggested a tall rear axle ratio. However, it all matched

Like AMC Concord AMX, Dodge Aspen R/T was a plush car given a sporty edge. That edge implied Rallye rims, E70-14 tires, and HD suspension; all standard. (Courtesy Eric Bremer)

The Aspen R/T and Volaré Road Runner represented the last time performance packs were offered on an A-body related car. (Courtesy Ronald Neuendorf)

With transverse torsion bar suspension, a 360-powered Aspen R/T could match a contemporary 'Vette in acceleration and through corners. (Courtesy Ronald Neuendorf)

At 198.8in long, and sitting on a 108.7in wheelbase, this 1977 Aspen R/T matched the rival Pontiac Trans Am. However, at around 3400lb, it was 300lb lighter. (Courtesy Ronald Neuendorf)

LA small block 360 V8 was good enough for the police force, and the R/T name! (Courtesy Ronald Neuendorf)

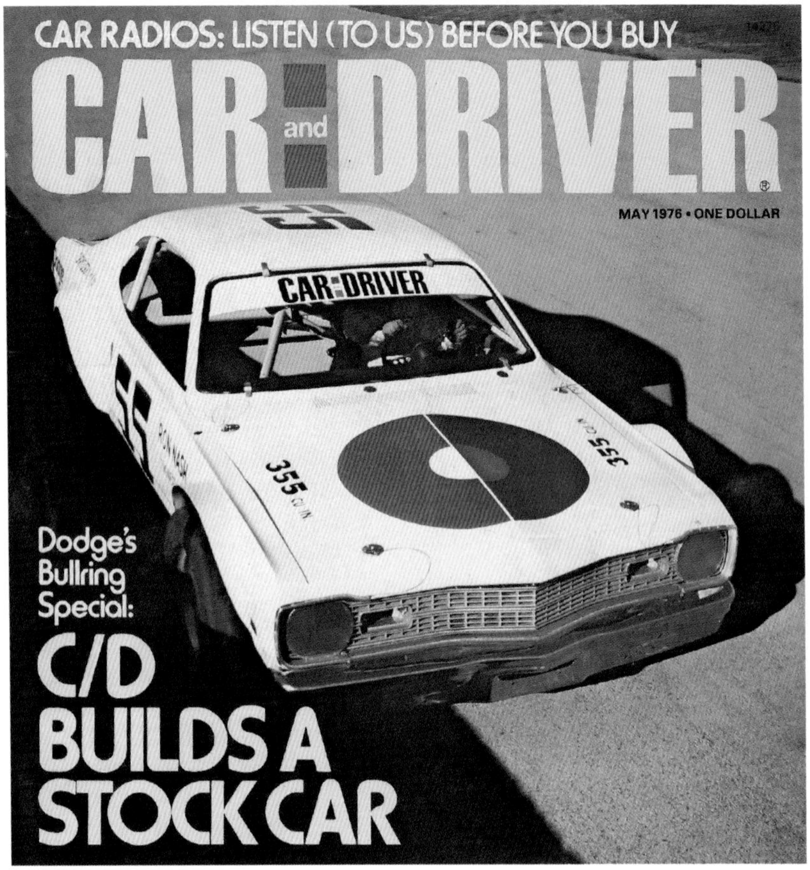

***Car and Driver*'s 1976 Dodge Dart 355 V8, built by Ron Nash of Roy Woods Racing. By 1975, Dodge's factory-supported 'Kit Car' stock car program had moved on to Dart Sport and Aspen. (Courtesy Car and Driver)**

the aforementioned super German's successor. *Car and Driver* tested a 40-plus grand Mercedes 450SEL 6.9, achieving high-15s and 140mph top speed. So an A43 coupe had an edge!

Sadly, in a commercial climate preoccupied with gas mileage and velvet interiors, the Super Coupe Kit Car was a tough sell. Aspen Super Coupe Kit Cars were painted two-tone red, with Plymouths in blue. Cars & Concepts of I96 Milford did the A43 package, with Direct Connection doing the 360's dress up parts.

Large spoilers, bolt on fender flares, and no hubcaps were pretty extreme ... or should that be Petty extreme? If exposed lug nuts weren't enough, then the No 43 door and roof decals would take one all the way to Daytona. There was even a 360 hood decal. The Kit Car was to honor NASCAR King Richard Petty's Mopar association. Trouble was, Petty had just switched from racing Dodge Magnum to the GM camp – as he had in 1969, with a brief Ford stay.

All those Kit Car decals were trunk delivered. However, A43 cars at the time tended to have the decals already affixed when the cars were on dealer lots. It seems that sales zone reps were able to convince dealers that Petty's GM switch was temporary, so the dealers put on the decals in hope. Unfortunately, it was hope in vain. Like all big buck sporty specials, the A43s stayed unsold for over a year. They even wound up at dealers in non-NASCAR country. No doubt the original southern dealers that took delivery just couldn't find buyers, so passed them along. With just 145 A43 Aspens and 247 A43 Volarés made, hindsight would have been a wonderful collector thing. A zero mile A43 coupe would be a winning 21st century ticket.

Concerning the factory's Kit Car program for short track racing, the mid '70s saw a model transition. With late model stock car racing, the factory made fewer than 12 Dodge Challengers in 1974. A switch was made to A-body Dart Sport, then Aspen, concerning steel tube chassis racers. Indeed, for ten grand over a stock F-body, the factory would supply a race car. Its tube chassis was created by Petty Enterprises. The car came as a bare rolling chassis, with or without bodywork. It could come as a turn-key racer.

The Fonz & Lee Iacocca say 'Sit on an F-body!'

In 1976, a one off show car called the Fonzmobile was created. The car was based on a Duster 360, and capitalized on the popularity of the TV show *Happy Days*. It featured a flame paint job, Lakes piped side exhaust system, faux dual spotlights, wide whites, baby moon hubcaps, fold down rear seat and Convertriple roof.

There was no intention to productionize Fonzmobile. It was all just great PR. However, some Californian Chrysler dealers got corporate blessing to do a California Custom. In this Mecca of car customizers, the California Custom could be Aspen or Volaré based. The Volaré edition used Aspen's hood and front bumper. Grille work was very SoCal, with a stainless steel tubular look. As usual, wide whites, baby moons/trim rings, and possibly functional Lakewood side exhaust made that Golden State 318 sound tough. More attention was paid to the greatly padded, Californian Carson-like top. No rear-quarter window, but lots of presence. At $1892.15, it was a custom job that cost one half the Volaré Custom Coupe's cost! Such was the price of individuality.

As horsepower went south, the custom car and van scene went into overdrive. The results weren't always pretty. With an eye to this custom/lifestyle movement, 1979 saw the factory Sport Wagon. With special stripes, front air dam, and fender flares, it was a R/T or Road Runner in hold-all guise. The van/truck scene also allowed one to get into looser commercial vehicle pollution and bumper laws, hence that iconic 360 V8 police pack powered Dodge Li'l Red Express of 1978-79.

1979 business saw Aspen R/T and Volaré Road Runner's décor pack have rear window louvers with horizontal slats. An 85mph speedo and engine diagnostic connector came, as Aspen and Volaré sales fell to 121,354 and 178,819 respectively. 1979 base prices for Aspen and Volaré were a respective $4018 and $4006. The R/T and RR's base 318 V8 had climbed to 155bhp in 1978, and declined to 135 horse in 1979. In 1980, the 360 was gone! The 318 fell to 120bhp at 3600rpm, but a four-barrel version was available with 155 horse at four grand, and 140bhp for California.

New Chrysler Corp boss Lee Iacocca felt F-body had been rushed into production. He insisted the family be restyled for its final 1980 model year, so it got hood, fenders, and front bumper from the F-body replacing M-body Dodge Diplomat. A thin-grille Fairmont visage, also; no doubt to annoy former corporate masters Ford. Even so, F-body was getting by on taxi and cop fleet car sales. 1980 numbers for Aspen and Volaré came in at 67,318 and 90,063, respectively. Just 285 Aspen R/Ts and 496 Volaré Road Runners rolled out in '80 MY. The end of Mopar muscle? Some thought so.

This modified '79 Plymouth Volaré Coupe has a 1985 AHB cop car 318, with 'Cuda 340 heads and intake manifold. It sports a '69 Road Runner six-pack hoodscoop, '78 Aspen grille, and rear 15x12in Slot Ansen rims. Shackles have been used to clear the big rear 325/50-15in M&H Drag Radials. (Courtesy Billy)

Chrysler Corp's troubled times

In the 1980 pilot episode of *MotorWeek*, John H Davis described the troubled state of the American auto industry. Sales by the domestics were 20 per cent down on August 1979's level. By contrast, imports were registering record sales. For the first time, they accounted for 30 per cent of the US auto market. Over one million workers in car making and related industries had been laid off. Plus, domestic automakers had surplus stock. With an industry average of 60 days supply, even Chrysler Corp's economical Dodge Omni and Plymouth Horizon subcompacts were at 90 days inventory. Ford Mustang and Mercury Capri were at 70 days, and even sensible Fairmont wasn't doing well, either. Oddly, for an import, there was a glut of the usually popular Datsun 280ZX.

It seemed slow sales were related to perceived quality/technology shortfalls, as well as the poorer gas mileage of domestic models. The post-second gas crunch conditions really favored the imports. Given such conditions, and a need to help the domestics catch up, Washington and Detroit called a truce. Proposed safety and tailpipe targets for 1983 were postponed. There was discussion on relaxing existing laws, like the 5mph impact bumper requirement. It was all a very long way from the Hemi V8, and F-body was in trouble.

In February 1980, *Consumer Reports* lined up four domestic compacts: AMC Concord, Chevy Citation, Dodge Aspen, and Ford Fairmont. Even with a TorqueFlite with torque converter lock up, Aspen would consume a projected 955 gallons of gas, compared to four-cylinder Citation's 655 gallons, over an annual 15,000 miles.[10] Unlike the other three test cars, Aspen didn't have a four-cylinder option. With gas prices being what they were, it explained why Citation had a waiting list and Dodge didn't. It was possible to pick up a pre-downsized domestic for a bargain price. Domestic dealers were also willing to deal favorably when talking turkey – so many were going out of business due to slow sales.

Buyers had to weigh-up lower price against poorer gas mileage, and resale value. 'You can pay me now, or pay me later' indeed. The Slant-6 had been a trusty friend, but its prewar ancestry was not helping efficiency. It had only four bearings, and bore spacing was uneven. Between cylinders two and three, plus four and five, it was 4in, compared to remainder spacing of 3.98in. Once upon a time it had been an inline-four. Indeed, in September 1957, *Road & Track* fondly recalled the 1931 PA Plymouth's smoothness, when judging the harshness of Volvo's HO B-16 I4.

In May 1975, *Road Test* had wondered how Chrysler Corp and others were going to bridge the economy and quality gap to the imports. The time for action was nigh. Chrysler had never been served well by its captive imports from Chrysler Europe. The French Simcas and British Plymouth Crickets had trouble against the VW Bug, then the Japanese imports.

Designing and making a small car in America had always been tricky; too little demand for economies of scale, too little mark-up to placate the UAW-bolstered labor costs. Washington had made the job harder. Smog controls caused domestic sixes and V8s to became thirstier, pushing buyers to imports. By the early '70s, VW had lost the econobox market to the Japanese, thanks to a stagnant Beetle. However, Wolfsburg was showing the way to the future with the North American-made Rabbit. Chrysler was on the right track, thanks to thrifty Omni and Horizon. In 1978, they became the first front drive domestic subcompacts made in America. It was belated fruit from Chrysler Europe. The Horizon was Simca's new family car, and showed Chrysler influence with its front torsion bar suspension. More than this, Walter P needed a homegrown compact for the eighties.

This magic Mopar wasn't going to come, or take inspiration, from an overseas subsidiary, because they were largely gone. Chrysler Europe had been sold for a token amount in 1978 to Peugeot, with shares for Chrysler and liabilities for the PSA Group. The French rebranded the cars as Talbots, and had dealings with Chrysler Australia at this time. The negotiations were to combine the big Talbot Tagora, originally slated as Chrysler Europe's next big rear driver, with Chrysler Australia's Hemi Slant-6. This car was going to replace the big, aging, Aussie-designed Valiant Downunder. In Europe, the Tagora would use four-cylinder power and the PRV V6. However, Chrysler Australia finished with its once popular Valiant, bought out by Mitsubishi in 1982.

Pennies from Washington

Now home alone, Chrysler USA's salvation would come from the K-car. In the spirit of Horizon, it was a transverse four-cylinder front driver, but first Walter P needed some money. Effectively bankrupt, and with a 1979 $1.7 billion loss, Iacocca's Chrysler arrival was a baptism of fire. He had to visit the Washington Overlords with the message that Chrysler was too big to fail. Uncle Sam said 'sure,' but first, a few restructuring conditions: Washington wished to see the K-car plans, 13 of Chrysler Corp's 47 plants were shut down, and better labor relations were a given. That meant workplace enterprise bargaining; taking shares over a pay rise, agreed layoffs, and other union concessions, all to achieve auto alchemy: a successful US automaker that majored in smaller, fuel-efficient cars.

History shows that K-car was a winner. More than that, the new design spawned the Dodge Caravan and Plymouth Voyager. These two models made minivans, the 1980s, and Chrysler Corp a family vehicle axis of economy. It seemed many needed no more than a turbo four – funny, how history repeats itself. For dynastic glamor, Mr Iacocca called on friend Alejandro De Tomaso. Chrysler had a 15 per cent stake in Maserati, and the two amigos cooked up a Caddy Allante rival in the front drive Chrysler-Maserati. Was there anything a Mopar turbo four couldn't do? If you were a traditional buyer, then yes.

The A-body rides again!

In 1983, Lee Iacocca and Chrysler Corp had gone almost completely front drive. Answering gas-guzzler criticism with concrete action. Iacocca noted their remaining V8 rear drive line was Chrysler's worst gas hog. They kept it for one reason: it yielded their hottest seller. Big Lee was of course talking about Chrysler M-body – the F-body's successor, or A-body version 3.0, if you will. Without this valiant performer, TJ Hooker would have had no wheels! In 1981, Aspen and Volaré gave way to the revised Dodge Diplomat, the Chrysler New Yorker and Fifth Avenue, and the Plymouth Gran Fury and luxo Imperial.

Chrysler Corp's formerly small cars became their upscale big cars. Four-cylinder front drivers like Dodge Aries and Plymouth Reliant were the economy 1960 Valiants for a new generation. Cops loved the Fifth Ave, and M-body was Chrysler Corp's last police car. It survived to 1989 on fleet sales. American luxury buyers also loved M-body, and the remaining big 'ol rear drivers from Ford and GM.

Car and Driver's Csaba Csere said today's front drive modern mobile rode more comfortably, and did a bunch of other things better, than Detroit's past efforts. Many Americans begged to differ. They felt a 3500-4000lb rear drive V8 gave a sense of security, comfort, and power that an econobox just didn't have. For them, Detroit kept traditional cars on the books – and very profitably.

Consumers had valid safety concerns regarding econobox security. Calspan did a 1979 offset crash test between an AMC Concord and VW Rabbit; it's no exaggeration to say Hornet ate Rabbit! The NHTSA's 35mph concrete barrier front impact test of 1979 Volaré Coupe was reassuring for owners. The Volaré's safety cell stayed intact, and its HIC driver reading of 724 was lower than the 800 recorded by a Mercedes 190E, even though the 190E had an airbag.

The only curious commercial misfire was the 1981-83 Imperial. This M-bodied luxury car, with smooth nose, fuel injected 318 V8, and Frank Sinatra ad campaign link, was Chrysler's take on the Nova to Caddy Seville transformation. It was probably a victim of the early '80s recession. Not even Iacocca's pal ol' Blue Eyes could save it. It was a relative failure Chrysler Corp could live with.

In 1982, Chrysler was making a $170 million profit. 1985 saw a record commercial performance as the corporation liquidated its Peugeot shares to the tune of $275 million, and yet Big Lee wasn't happy. Chrysler was very reliant on its K-car, and lacked product diversity. Even engine options saw a Grand Canyon between the turbo fours, and 318/360 small block V8s. Many Americans wished for a V6. The Dodge truck and pickup areas were still solid – so, too, the captive import connection with Mitsubishi. However, Chrysler needed more; it needed Jeep.

Highland Park swallows Kenosha

AMC's 1980s alliance with Renault hadn't gone to plan. The idea was for Renault to become a North American force in the manufacture and sale of I4 front drive econoboxes, while AMC concentrated on Jeep. The trouble was that the Alliance and Encore didn't deliver on quality. As gas prices fell, so did Alliance and Encore sales. The core econobox market still belonged to the Japanese. That market wasn't big enough to sustain any more players until the Koreans came. In the meantime, Kenosha was a little sleepy. In 1986, it was contracted to make Omnis, Horizons and Fifth Avenues. Also in 1986 came the formal announcement that Chrysler Corp would be taking over AMC.

It was a deal worth making. In 1988, the XJ Cherokee accounted for one third of Chrysler's profits! Under the buyout deal, Chrysler was obligated to sell some Renaults in America. Of course Chrysler wanted AMC for Jeep, not Alliance or Premier, and preferred Mitsubishi for captive imports. Iacocca went back on a promise to retain the AMC brand, going with Eagle for Kenosha refugees and captive imports. Even so, American Motors exerted influence in the Mopar family. The North-South engine layout front drive Eagle Premier influenced development of Chrysler's LH Taurus fighters.

Ex-AMC staff oversaw development of the Neon. Chrysler's Director of Small Car Operations was also from AMC. It also had a hand in the PT Cruiser. Francois Castaing had moved from Renault to AMC. With the Renault F1 turbo V6 and XJ Cherokee in his portfolio, he was chief engineer on team Viper. More than this, AMC got Chrysler out of the Ford/GM style of rigid departments into more free-flow work groups. Cab forward design, Viper, Prowler, and PT Cruiser; by the 1990s, Chrysler Corp had certainly gone beyond K-car. It was the funkiest member of the Big 3, as GM fought for early '90s economic survival, and Ford decided to put its oval grille theme outside and inside its cars!

The slings and arrows of outrageous business

Sadly, individuality isn't always commercially rewarded. It is said that CEO Bob Eaton's lack of forward planning made Chrysler Corp desirable as a takeover target; or maybe it was just Jeep? Either way, that super German Daimler-Benz came knocking on Chrysler's door. The result was the 1998 'Merger of Equals,' and Mopar devotees having to get used to saying 'Daimler-Chrysler.' In the early '90s auto world, there was this notion that a company had to get super big, or die. So, Mercedes got Chrysler, and BMW bought Rover; temporary marriages, in both cases.

It wasn't really a merger of equals – more like a real culture clash. Chrysler management were given 'etiquette booklets' to ease discourse with their new Swabian brethren. On the plus side, some Chrysler Corp cars got hold of Mercedes hardware to help transition back to rear drive cars with Hemi V8 power. Ultimately, Daimler-Benz sold Chrysler to Cerberus Capital Management in 2007. Precipitated by the GFC. The subsequent bankruptcy, and taxpayer bailout under TARP (Troubled Asset Relief Program) opened the door to 'Fix It Again Tony' (Fiat).

Fiat had left North America just a little before Renault, and for similar

Car customizer Carl Green, with his CGE Pacer Pickup of the late '70s. When doing 1990s Chrysler concept cars, Green encountered many AMC folk from the old days. (Courtesy Carl Green)

Marie and Erika of pop classical duo MARIERIKA, with the retro Fiat 500. The modern 500 was the idea of FCA CEO Sergio Marchionne. (Courtesy twitter.com/marierika_2010)

reasons – basically, an inability to mix it with the Japanese on quality. Now they were back, but fortunately without Malcolm Bricklin. Canadian-Italian accountant Sergio Marchionne was now the boss of Fiat. He had turned around the Turin titan's fortunes with value added transformations. Take one basic Fiat Punto, dress it up, and sell more expensively as the Fiat 500 retro ride and Alfa Mito. The Chrysler faithful were hoping for the second coming of Lee Iacocca.

Since 2014, Marchionne has held

The 1987 AMC buyout got Chrysler the Jeep brand. It also brought creative input, to get Walter P beyond minivans and K-cars. Automatic Pacers utilized the Chrysler TorqueFlite A904. (Courtesy Carl Green)

FCA (Fiat Chrysler Automobiles) together through difficult times, but has faced criticism. It has been along the lines of 'long on promise.' Mopar fans have considered his actions as raiding the Chrysler/Jeep piggy bank to finance the revival of Alfa Romeo and Maserati at a time when the home designed product could have used modernization money. But then again, who doesn't enjoy a good revival?

The LX platform road to Hemi heaven

In the 1963 December issue of *Esquire*, Virgil Exner showed his Revival Cars; These were stylistic re-interpretations of how hallowed names from America's auto past would look today. The cars shown were Stutz, Duesenberg, Packard, and Mercer. Exner went on to create more Revival themed cars, which were made into 1:25 scale plastic model kits by company Renwal. Chrysler itself was started from the ruins of another iconic American company – the Maxwell Motor Co. On his radio and TV show, comedian Jack Benny owned a Maxwell. Benny's thrifty persona overcame the Maxwell's broken steering by getting Rochester to kick the wheel when they came to a corner! Fortunately, Chrysler Corp isn't consigned to history, and the concern really turned the corner when it revived the 300C in 2004.

The new 300 – letters of distinction

The 21st century Chrysler 300 took its inspiration from the classic Letter Series cars of the '50s, as well as the Monteverdi cars of the '70s. In both cases, luxury was combined with Hemi V8 power, with the Swiss connection bringing four doors. The modern 300's genesis started with the Concept 300 of 1991. This Viper-engined dream machine took inspiration from Monteverdi's High Speed GT machines. The idea that sporty cars could have four doors was a European thing. The international-natured Chrysler 300 is one of the few American cars in history that is factory designed for left and right-hand drive markets.

Built at Brampton Assembly, Canada; Graz, Austria, by Magna Steyr; and in Beijing China, by Beijing Benz; the 300 has something in common with the Mini – whereas the original was mostly known only at home, the modern version is commercially successful all over the world. The first generation 300, running from 2004 to 2010, stemmed from a 2001 Ralph Gilles design. As Gilles reflected upon in 2008, it became a latter day Chrysler icon. Next stop was concept car duty at the 2003 New York Auto Show. US sales started in April 2004.

As the first full-size Chrysler Corp rear driver since the 1981 Chrysler Newport, the 300 was traditional in every way. Not just a big rear drive car with optional V8 punch, but selling at those old Motown volumes – the 300 family managed over 110,000 units per annum for its first four years, and over 50,000 sales per year between 2012 and 2015. Plus, it wasn't an SUV! Like those revered inters and full-sizers of yore, the bidding started with six cylinders – the Pentastar V6 to be precise – but dreams were made of V8s.

A development of the flexible LH platform, the LX platform made rear drive and Hemi's comeback a reality. The latest design accommodated Mercedes hardware from the Daimler-Chrysler tie up. This encompassed the Mercedes W211 E-class multi-link rear suspension, front seat frames, wiring harness, steering column, five-speed autobox, and eventually a version of MB's 4Matic AWD system for the snowbelt crowd. Two initial star powerplants were the 340 horse Hemi 5.7 – yes, a real Hemi – and the Hemi 6.1L slayer.

Efficiency came from an induction tube on the engine block's side that pulled air, and pushed it into the induction area; shades of Ram Induction and '60s Hyper-Pak. The dual sparkplugs per cylinder recalled the '70s development work Mopar engineers did

A Gremlin X on the strip! The 1987 AMC buyout brought a quirky element to Chrysler Corp. (Marc Cranswick)

with Dick Landy Incorporated concerning twin-plug Hemi heads. A Multi-Displacement System (MDS) allowed owners to have their cake and eat it, with EPA numbers of 15mpg city/23mpg highway for five-speed auto. MDS dropped the V8 to four-cylinder mode, when the going was easy. The Hemi 6.1 passed on MDS.

While all versions carried a big grille reminiscent of the '50s Letter Series rocket sleds, only the upscale 300 was denoted 300C. The hot 300SRT8 (Street & Racing Technology) was a budget AMG/M5 rival of surprising ability. Big, but effective, and rather raw, all invoking those good 'ol '60s muscle car days. The 300SRT8 came with all the fruit, like 20in rims, wheelarch spats, chrome mesh grille, MyGIG sat nav, SRT steering wheel, SRT8 tooled leather sports seats, and carbon fiber interior trim. That last item replaced the 300C's plastic Tortoise Shell faux wood trim, which itself was visually pleasing.

Motor Trend was pleased with 300C, naming it 2005 *Motor Trend* Car of the Year. *Car and Driver* had it in its 'Ten Best' list in 2005 and 2006. Even Clarkson and crew of BBC's *Top Gear* were welcoming of 300C's value for money – Jeremy Clarkson even using the standard power seat to elevate his head out of the sunroof, not unlike a Dave 'Big Deal' model kit driver. In Britain, the original LX 300C was seen as the poor man's Bentley. Some even swapped Bentley badges onto their sedans!

In America, the 300C was judged

seriously 'gangsta.' A factory pimped-out ride that became a hip-hop, rap music video regular, with Snoop Dogg, in particular, being one patron. A 300C fan, Snoop Dogg contacted Mercedes man and then-Chrysler CEO Dieter Zetsche in 2004, requesting a personalized 300C. Snoop Dogg later appeared in a TV commercial with Lee Iacocca. Chrysler icons of the past and present for sure. In spite of the new 300's global nature, Chrysler's commercials played on the Detroit connection.

The 2011 'Imported From Detroit' campaign, featuring Eminem, debuted at the Super Bowl. In 2012, Berry Gordy Jr did a TV ad saying "We are Motown, and this is what we do," all to the tune *Ain't No Mountain High Enough*. The same year, a limited run of 500 Mopar '12s were created by Mopar Performance to commemorate Mopar's 75th Anniversary. It featured a short 3.91 rear axle, sports suspension, and unique badging.

Then came the 2013 '300 Motown Edition,' with special chrome rims, Beats audio, 'Motown Edition' front fender badges, and 100 Motown tunes preloaded on the USB drive – so there will always be a place where Ike and Tina get along! In modern times, Mopar Performance is a subdivision making high-performance aftermarket parts for Chrysler Corp rides.

In truth, the Detroit connection was exaggerated: the Hemi V8 is built in Chrysler's Mexican plant, the North American first gen 300 was Canadian built, and Chrysler's Jefferson North Assembly was the only car plant in Detroit. However, even Motown Records moved to California. After that front drive, V6-powered 300M of the '90s, Mopar fans were grateful and relieved by the LX platform 300's return.

In 2011, the second generation Chrysler 300 made its debut; less gangsta, but with more high-tech, and a smaller, more discreet grille; fewer Mercedes underpinnings, but greater use of expensive high-strength steel for thinner pillars, and a more steeply raked windshield. Bi-Xenon HID projector headlights, LED daytime running lights, LED taillights, and an eight-speed ZF 8HP 45 auto brought the 300 up-to-date. There was leather for all trim levels, and a modernized take on the 1930s Chrysler winged emblem. The eight-speed applied to the V6 version from 2012 MY; the five-speed 'Auto-Stick' 5G-Tronic automatic to the 3-liter V6 turbo diesel, and 6.1-liter replacing 6.4-liter Hemi V8 – the latter now with miserly MDS.

The 300's European diesel option had always come from Italian diesel specialist VM Motori. Between 2011 and 2014, CEO Marchionne used this motor and the second generation 300 to revive the Lancia brand in Europe, with the reborn Lancia Thema. Magna Steyr's work assembling European delivered and right-hand drive 300s stopped with the first gen 300, so the Thema was built at Brampton, Ontario! To avoid model duplication with the LD platform Charger and LC Challenger, the US 300SRT8 ended in 2014. However, it continues in export markets, like Australia and Japan, that don't get Charger or Challenger.

Australia also still receives the decontented 6.4-liter 300SRT Core, a value-for-money stripper that passes on some of the normal SRT8's tricks. Whether regular or Core, a 6.4-liter Hemi and 470lb/ft of torque spelt 0-60mph in a tad over four seconds. Special lower-front fascia, oversized exhaust outlets, body color in place of chrome, 20in alloys, lowered suspension, and Brembo brake package were all parts of SRT8's bag. To paraphrase MC Hammer, you can't touch *that* for value.

For those that don't need the whole SRT experience, there's the sports flavored 300S. This iteration arrived at the 2011 New York International Auto Show, and has blackout treatment for grille headlamps, 20in polished face alloys with black painted pockets, ten speaker 'Beats by Dr Dre' sound system, and steering wheel mounted paddle shifters. Whichever Chrysler 300 chosen – for sports, luxury, or both – it delivers that traditional big car, rear drive feel unique to America; and interior materials were classier the second time around.

Got Hemi? ... Yup!

It was an answer Mopar fans had been waiting 32 years to hear, and it came in a memorable 2003 TV commercial for the Dodge Ram Heavy Duty Truck. The truck is towing a classic B-body Charger, when a couple of fellas pull up in a Duster to ask: "That thing got a Hemi?" The trucker answers "yes," and takes off with great alacrity. The Duster boys are surprised by the rig's turn of speed. At the next set of lights the trucker adds, "by the way, the Charger has a Hemi too!"

So it was, that this TV ad heralded the return of Hemi V8 power to Mopar land for the first time since 1971. The ad promoted the Ram HD availability of a 5.7-liter Magnum Hemi V8. Optionable in Dodge Ram 1500, 2500 and 3500 pickups. This complemented the 245bhp Magnum 5.9, aka ye olde 360 LA V8. It was certainly worth the wait, and it was certainly a Hemi – even if the heads were flatter this time around. With distributorless ignition, and aforementioned dual sparkplugs per cylinder, the 345in^3 modern Hemi hit one horse per cube; however, make that one net horsepower, not gross! The MDS mill had a handy 398lb/ft to go with its traditional OHV layout. Sorry Henry, no overhead cams here.

The new Hemi V8 was made exclusively at Chrysler's Saltillo engine plant in Ramos Arizpe, Mexico. With a 3.917in bore and 3.578in stroke, the 5654cc motor promised more scat than Speedy Gonzalez! As with all MDS modern Hemis, it could switch to two cylinders per bank operation on light throttle to save gas. Unlike in the old Caddy V8-6-4 days, the MDS was more economical and reliable. In six-speed manuals with HD trucks, non-MDS Hemi V8s lived. Same deal for the 2009 Dodge Challenger R/T.

Beyond trucks, the new Hemi could obviously be used under the hoods of Chrysler Corp cars. So, apart from Dodge Durango and Grand Cherokee, the Hemi 5.7 could help the 300C, Magnum R/T,

MARIERIKA with the 2017 Chrysler 300S. (Courtesy www016.upp.so-net.ne.jp/MARIERIKA/)

Since 2010, Brampton Assembly (Canada) has taken over the building of right-hand drive 300s from Austria's Magna Steyr. (Courtesy ja-jp.facebook.com/marierika2006/)

The LX platform Chrysler 300S, shown here at Fujisawa Shonan Chrysler-Jeep, is a sports-trimmed version of the 300 family. (Courtesy youtu.be/surtdwQcPY8)

Charger R/T and Challenger R/T. Modern Mopar muscle cars had arrived! In 2009, the Hemi got variable valve timing – VVT in Chrysler's literature. An oil control valve sent oil to a special cam sprocket to advance or retard timing. At this hour, better flowing heads joined an active intake manifold on non-hybrid vehicles. A valve was used to switch from long runners (torque), to short runners that favored power, all based on RPM.

The first super-hot new Hemi V8, was the 2005-2010 Hemi 6.1. This motor had a different block compared to Hemi 5.7, with revised coolant channels and oil jets to spray cool the pistons. A 4.06in bore and same 5.7-liter stroke made for 6059cc or 370in^3. It made a classic Hemi 426 matching 425 horse, but net rather than gross, and at 6200rpm, with 420lb/ft at 4800rpm. This big baby could rev! Just as well the Mopar Magicians added a forged crank, lighter pistons, and stronger conrods. The Hemi 6.1 V8 was set up for top-end power, so no variable runner length for the aluminum intake breathing straw, and hold the MDS.

On 2005's show circuit, a 392in^3 Apache 6.4-liter Hemi V8 was showcased. Apache represented the state of the art. With high strength forged alloy pistons, this V8 made a conservative 525 horse and 510lb/ft. Once again, they were all net figures. The Apache made a 2007 crate motor debut as the 392 Hemi. It invoked a hallowed displacement figure from the '60s. Don Garlits used a few! Confusingly, in 2011 the Hemi 6.1 production engine was replaced by the Hemi 6.4, which was also a 392in^3 V8. However, this 392 was based on the Eagle 5.7 Hemi, and had little in common with the 392 crate motor.

By 2011, the Chrysler 300C's 5.7 had moved up from its 2004 numbers of 340bhp and 390lb/ft to 363 horse and 394lb/ft. With the 2014 MDS Hemi 5.7 at 395 horses. The Hemi 6.4 V8 had variable valve timing and MDS. With the best of both worlds, it made 470bhp and 470lb/ft, but MDS meant it didn't fear the stricter CAFE requirement President Obama brought, in the wake of soaring 2010 gas prices.

In 2015, the 21st century Hemi reached a new zenith, with the Dodge Charger and Challenger Hellcats. These ultra supercars had 707 horses and 650lb/ft, and were named after the Grumman F6F Hellcat. Both cars used a supercharged 6.2-liter motor, with 9.5:1 CR and no MDS. They were the most powerful muscle car engines to that moment in time. Specifications involved a Hemi 6.4 bore of 3.578in, and Hemi 5.7 stroke. The 6166cc or 376in^3 V8 had a twin-screw 145in^3 IHI supercharger, with integrated charge coolers, and 11.6 pounds of max boost. You want Hemi? You've got Hemi!

The General Lee returns – LX, LD Charger

From 1975, the B-body Charger went into the sports/luxo personal coupe market, with the Magnum as a buddy in arms, plus paint-and-tape Daytona too. In the '80s, Charger came back as a front drive I4 Omni relative. Carroll Shelby came over, and the 1983 Shelby Charger, Omni GLHS, and Dodge Daytona Turbo Z CS were all competent cars. The Omni GLHS (Goes Like Hell Some-more) became a turbo four icon. In independent testing, it outperformed a '60s Shelby Mustang GT350 in acceleration, braking, lap times and naturally, fuel economy. However, enthusiasts kept thinking back to Waylon Jenning's little song about a couple of boys never meaning no harm.

In 2006, the Dodge Charger returned as an LX platform based four-door car. Yes, that's right! Four doors – no waiting to be picked up. So, unlike a certain orange colored '69 Charger with a flag on the roof, there was no need to jump over welded doors. This here new moonshine special had room for the whole Duke clan. Bo, Luke, Daisy, and Uncle Jesse could enjoy the ride. Cooter might have had a little trouble with the

A 2014 Dodge Charger Enforcer AWD police fleet car. Commercially, an over-reliance on fleet sales hasn't always helped Chrysler Corp's profitability. (Courtesy Lee Rehorn)

The LD Charger's 2011 makeover saw a high-quality interior, and continued use of the five-speed automatic column shift. (Courtesy Lee Rehorn)

The modern Hemi V8, in 5.7 liter (345in³) MDS VVT form. The new Hemi 5.7 V8 started at 340 horses in the 2004 Chrysler 300C. (Courtesy Lee Rehorn)

Who would have thought the new General Lee would be used as a cop car? Certainly not Boss Hogg! (Courtesy Lee Rehorn)

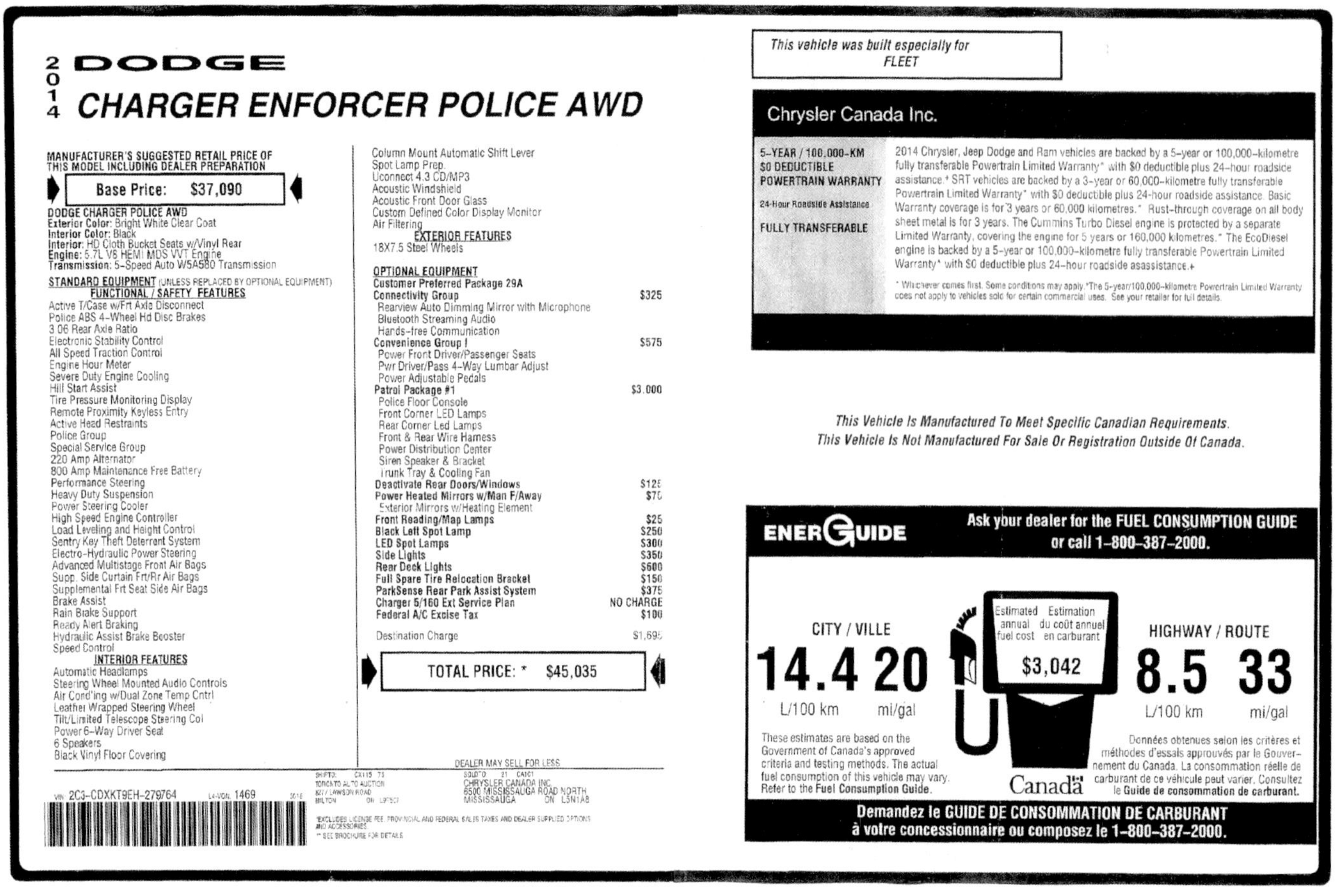
2014 DODGE
CHARGER ENFORCER POLICE AWD

MANUFACTURER'S SUGGESTED RETAIL PRICE OF THIS MODEL INCLUDING DEALER PREPARATION

Base Price: $37,090

DODGE CHARGER POLICE AWD
Exterior Color: Bright White Clear Coat
Interior Color: Black
Interior: HD Cloth Bucket Seats w/Vinyl Rear
Engine: 5.7L V8 HEMI MDS VVT Engine
Transmission: 5-Speed Auto W5A580 Transmission

STANDARD EQUIPMENT (UNLESS REPLACED BY OPTIONAL EQUIPMENT)

FUNCTIONAL / SAFETY FEATURES

Active T/Case w/Frt Axle Disconnect
Police ABS 4-Wheel Hd Disc Brakes
3.06 Rear Axle Ratio
Electronic Stability Control
All Speed Traction Control
Engine Hour Meter
Severe Duty Engine Cooling
Hill Start Assist
Tire Pressure Monitoring Display
Remote Proximity Keyless Entry
Active Head Restraints
Police Group
Special Service Group
220 Amp Alternator
800 Amp Maintenance Free Battery
Performance Steering
Heavy Duty Suspension
Power Steering Cooler
High Speed Engine Controller
Load Leveling and Height Control
Sentry Key Theft Deterrent System
Electro-Hydraulic Power Steering
Advanced Multistage Front Air Bags
Supp. Side Curtain Frt/Rr Air Bags
Supplemental Frt Seat Side Air Bags
Brake Assist
Rain Brake Support
Ready Alert Braking
Hydraulic Assist Brake Booster
Speed Control

INTERIOR FEATURES

Automatic Headlamps
Steering Wheel Mounted Audio Controls
Air Cond'ing w/Dual Zone Temp Cntrl
Leather Wrapped Steering Wheel
Tilt/Limited Telescope Steering Col
Power 6-Way Driver Seat
6 Speakers
Black Vinyl Floor Covering
Column Mount Automatic Shift Lever
Spot Lamp Prep.
Uconnect 4.3 CD/MP3
Acoustic Windshield
Acoustic Front Door Glass
Custom Defined Color Display Monitor
Air Filtering

EXTERIOR FEATURES

18X7.5 Steel Wheels

OPTIONAL EQUIPMENT

Customer Preferred Package 29A	
Connectivity Group	$325
Rearview Auto Dimming Mirror with Microphone	
Bluetooth Streaming Audio	
Hands-free Communication	
Convenience Group I	$575
Power Front Driver/Passenger Seats	
Pwr Driver/Pass 4-Way Lumbar Adjust	
Power Adjustable Pedals	
Patrol Package #1	$3.000
Police Floor Console	
Front Corner LED Lamps	
Rear Corner Led Lamps	
Front & Rear Wire Harness	
Power Distribution Center	
Siren Speaker & Bracket	
Trunk Tray & Cooling Fan	
Deactivate Rear Doors/Windows	$125
Power Heated Mirrors w/Man F/Away	$70
Exterior Mirrors w/Heating Element	
Front Reading/Map Lamps	$25
Black Left Spot Lamp	$250
LED Spot Lamps	$300
Side Lights	$350
Rear Deck Lights	$600
Full Spare Tire Relocation Bracket	$150
ParkSense Rear Park Assist System	$375
Charger 5/160 Ext Service Plan	NO CHARGE
Federal A/C Excise Tax	$100
Destination Charge	$1,695

TOTAL PRICE: * $45,035

DEALER MAY SELL FOR LESS.

VIN: 2C3-CDXKT9EH-279764 L4-VON: 1469

SHIP TO: CX115 73 TORONTO AUTO AUCTION 8277 LAWSON ROAD MILTON ON L9T5C7

SOLD TO: 21 CA101 CHRYSLER CANADA INC. 6500 MISSISSAUGA ROAD NORTH MISSISSAUGA ON L5N1A8

*EXCLUDES LICENSE FEE, PROVINCIAL AND FEDERAL SALES TAXES AND DEALER SUPPLIED OPTIONS AND ACCESSORIES.
** SEE BROCHURE FOR DETAILS

This vehicle was built especially for FLEET

Chrysler Canada Inc.

5-YEAR / 100,000-KM
$0 DEDUCTIBLE
POWERTRAIN WARRANTY
24-Hour Roadside Assistance
FULLY TRANSFERABLE

2014 Chrysler, Jeep Dodge and Ram vehicles are backed by a 5-year or 100,000-kilometre fully transferable Powertrain Limited Warranty* with $0 deductible plus 24-hour roadside assistance.+ SRT vehicles are backed by a 3-year or 60,000-kilometre fully transferable Powertrain Limited Warranty* with $0 deductible plus 24-hour roadside assistance. Basic Warranty coverage is for 3 years or 60,000 kilometres.* Rust-through coverage on all body sheet metal is for 3 years. The Cummins Turbo Diesel engine is protected by a separate Limited Warranty, covering the engine for 5 years or 160,000 kilometres.* The EcoDiesel engine is backed by a 5-year or 100,000-kilometre fully transferable Powertrain Limited Warranty* with $0 deductible plus 24-hour roadside asassistance.+

* Whichever comes first. Some conditions may apply. +The 5-year/100,000-kilometre Powertrain Limited Warranty does not apply to vehicles sold for certain commercial uses. See your retailer for full details.

*This Vehicle Is Manufactured To Meet Specific Canadian Requirements.
This Vehicle Is Not Manufactured For Sale Or Registration Outside Of Canada.*

ENERGUIDE

Ask your dealer for the FUEL CONSUMPTION GUIDE or call 1-800-387-2000.

CITY / VILLE
14.4 L/100 km 20 mi/gal

Estimated annual fuel cost / Estimation du coût annuel en carburant
$3,042

HIGHWAY / ROUTE
8.5 L/100 km 33 mi/gal

These estimates are based on the Government of Canada's approved criteria and testing methods. The actual fuel consumption of this vehicle may vary. Refer to the Fuel Consumption Guide.

Données obtenues selon les critères et méthodes d'essais approuvés par le Gouvernement du Canada. La consommation réelle de carburant de ce véhicule peut varier. Consultez le Guide de consommation de carburant.

Canadä

Demandez le GUIDE DE CONSOMMATION DE CARBURANT à votre concessionnaire ou composez le 1-800-387-2000.

Domestic car fans welcomed Chrysler Corp's return to V8 rear drivers. The excellent gas mileage of LX family cars would have been 1960s sci-fi stuff! (Courtesy Lee Rehorn)

new fangled Hemi V8 after that good ol' Magnum 440. The Dukes might have passed on painting '01' on the doors, to keep a low profile from the law.

With everything old being new again, retro touches on the new Charger involved a B-body reminiscent hood stamping, profile swage line, and taillight treatment. The initial top model was Charger R/T with MDS Hemi 5.7 and five-speed Autostick. The first gen LX Charger also had a SRT8 variant, with 6.1-liter V8. Because modern muscle includes luxury, Charger SRT8 brought eight-way powered, heated seats, with perforated suede inserts, powered and adjustable pedals, auto climate control, and body-color matching interior trim. There was a three-person memory function for the seat/pedal positions. GPS sat nav and

A Hurst Heritage GSS Charger. From 2008, Mr Norm added his name to modern high-performance Mopars. (Courtesy Larry Weiner, mrnorms.com)

322 watt audio system came along for the ride. Outside touches saw a custom grille, special front fascia, rear spoiler, SRT8 color choices, and unique trim. Underhood inspection would reveal an SRT8 engine cover. Higher feel steering gear, and larger exhaust outlets were all to the good. An optional Road/Track pack brought even more good stuff: an extra 100bhp, GPS sat nav, 322 watt audio, rear seat DVD entertainment system, and a sunroof!

In 2011, the reworked LD Charger appeared, with a higher quality interior, and more exterior retro touches. The exterior styling drew upon B-body Charger's most beloved 1968-70 era. Therefore, expect sculpted profile door panels; hello General Lee, and rear wraparound LED taillights. The latter would be familiar to anyone that has seen the 1968 movie *Bullitt*. On a practical level, there was 15 per cent more greenhouse. This answered complaints concerning outward visibility – a problem with many modern cars, especially sporty ones.

Muscle cars, perhaps; but since 2010, many sporty car buyers have been watching the pennies. There's rising interest in sporty value from non-V8 versions. So, LD's change from the old 3.5 V6, to the revised Pentastar 3.6 V6, was welcome. In budget performance world, replacement of the four-speed auto with five-speed Autostick was a shift in the right direction. Charger SRT8 wasn't around in 2011, but returned for 2012. It was the only version not available with AWD.

2012 saw yet another fabled Mopar arise once more: the Dodge Charger Super Bee; special badging, Hemi 6.4 470 horse mill, four-piston slotted rotor calliper Brembos, paddle shifters, custom seat embroidery, SRT8 launch facility, and in-dash, on-screen performance telemetry – popular on modern, fast cars, such info involved readouts for 0-60mph, live lateral G-force figure and quarter-mile time. A further restyle in 2015 occurred, using LED lights to dazzling effect. Modernizing brought an eight-speed auto, but AWD was just for Charger V6, now. In addition to 5.7-liter R/T and 6.4-liter SRT8, the Charger Hellcat made for an uncommon range of family cars.

***Pentastar Power* magazine editor Larry Weiner looks after vehicle and parts supply concerning the program of modern Mr Norm vehicles. (Courtesy Larry Weiner, mrnorms.com)**

The 707 horse supercharged 2015 Dodge Challenger SRT Hellcat established a new standard in reasonably priced, high-performance cars. (Courtesy Bob Curry)

Wagons roll! LX Magnum

With a shot at the once popular family wagon market, the Dodge Magnum was an LX platform relative of the Chrysler 300 family, built in Brampton, Ontario for North American sales, and assembled by Austrian company Magna Steyr concerning export sales. In Europe and Australia, the Dodge Magnum was sold as the Chrysler 300 Touring. This export edition featured the popular 300C front fascia and luxury interior. An export SRT8 version of this Touring was also available. It made for an excellent Shooting Brake.

The Magnum was a critical success, making *Car and Driver*'s Ten Best list on debut in 2005. Running from 2005 to 2008, Magnum sales were in the 30,000 to 50,000 unit range. The once-mighty suburban standby, the American wagon, was now sadly a bit player in an SUV movie. Hot versions were the 340bhp Hemi 5.7 R/T, and 425 horse Hemi 6.1 motorvated SRT8. Both used the Mercedes W5 A580 five-speed auto, and the Magnum 5.7 R/T could have AWD. The ultimate niche product!

The Magnum SRT8 started as a 2003 LA Auto Show concept car. It made its debut at the 2005 same show, and went on sale in 2006 model year. Running on 20in rims, possessing Brembo brakes, firmer sports suspension, special lower body treatment, and custom front and rear fascias, the SRT8 was no wallflower, and no shrinking violet either. Its performance was loud and proud. In March 2007, *Motor Trend* recorded 0-60mph in 5.1 seconds, 0-100mph in 11.7 seconds and a 13.1 second quarter-mile at 108mph. That's some fancy shooting, partner!

Sadly, the Magnum SRT8's call to arms fell on deaf ears. 2970 Magnum SRT8s were made in 2006, 921 in 2007 and just 239 in 2008. Blame it on the SUV. In 2008, the Magnum was

The twin-screw 145in^3 IHI supercharger delivers a maximum of 11.6psi to the Hemi 6.2 liter V8 (376in^3). At introduction, the Hellcat motor was the most powerful muscle car engine ever created! (Courtesy Bob Curry)

facelifted inside and out, matching Chrysler 300's changes. The SRT8 got a new hoodscoop, and all-new bright red exterior paintjob, where color keying even covered the bumpers. However, on November 1, 2007, Chrysler Corp announced that Magnum, plus PT Cruiser convertible, Crossfire, and Pacifica were not profitable, so all were deep sixed; and with them the last US midsize wagon, it would seem.

LC Challenger – close, but no 'Cuda

In the 1985 *TJ Hooker* episode 'Taps for officer Remy,' Sergeant Hooker hunts down a cop killer, using a tire print from the crime scene. A plaster cast is made from the print to determine whether a light or heavy vehicle is at hand. The statement is made to look for high-performance tires for a car above 2500lb; that is to say, in those Chrysler cop car M-body days, a vehicle weighing 2500lb or less was usually an econobox – such a car couldn't be a high-performance vehicle, needing high speed tires.

Fast-forward 30 years, and the high-performance world the new LC Dodge Challenger was about to enter was very different. The sport compact label had been hijacked by Japanese four-cylinder imports. A '70s CVCC Civic was no hot rod, but a nitrous sniffing VTEC Civic might get a part in a *Fast & Furious* flick. Similarly, Mustangs and Camaros had changed. Turbo four options that actually worked, and handling packs that made them track day cars? You can't get a four-banger in a Challenger, and you wouldn't want one: too many bad '80s memories of freeway on ramp straining Iron Dukes.

It took some tricky broken-field running to get Challenger's nameplate back to V8 glory from a one-time Mitsubishi captive import. According to *Autoweek*, the Challenger SRT8 392, fits in closer with traditional muscle car folklore. It's a car you want to do smoky burnouts in on a deserted street in the dead of night. That said, the Hellcat

As with all high-echelon modern Mopars, Hellcats use a Brembo brake package. This helped one Challenger SRT Hellcat to lap the Nürburgring in 7 minutes and 50 seconds. (Courtesy Bob Curry)

iteration has lapped the Nürburgring in 7 minutes and 50 seconds, so it's not all straight-line thinking. The Dodge Challenger concept dated back to a 2006 design done by Jeff Gale and Brian Nielander. It was displayed at the Detroit Motor Show that year. As with other reborn Mopar Heroes, it can trace engineering lineage to the LH platform.

Like retro fifth gen Camaro, Challenger had a concealed B-pillar for that traditional hardtop look. It also sported DaimlerChrysler derived hardware. Compared to Chrysler 300 and Dodge Charger, Challenger

In the old days, the SRT Hellcat would have been described as coming with 'His & Her' key fobs. The black key fob was good for 500bhp; the red one unlocked the stable's full 707 horses! (Courtesy Bob Curry)

Stock Challenger SRT Hellcat performance involved a 0.94g skidpad figure, and top speed exceeding 200mph! (Courtesy Bob Curry)

lived on a truncated platform. It had Mercedes W220 S-class control arm front suspension, W211 E-class five-link rear suspension, along with W5A 580 automatic, rear differential, and ESP. Regarding dimensions, Challenger was 197.7in long, 75.7in wide and 57in high. Sitting on a 116in wheelbase, a V6 Pentastar-powered SXT weighed in at a lithe 3834lb, whereas SRT Hellcat S/C comes in at 4469lb. As for the six-speed Tremec TR 6060, like the LC Challenger, there was nothing like this back when the E-body was current.

On street-legal drag rubber, Hellcat could scat to sixty in 3.6 seconds, and stop from 60mph in 109ft. This car's specs involve Ivory Pearl A8 paint, red seat belts, and sunroof. It was delivered on September 16 2015. (Courtesy Bob Curry)

Spec-wise, the range used to start with Challenger SE, then came mid-level R/T. The R/T had Hemi 5.7 and five-speed auto standard. This sensible combo came with MDS and could have iDFSO (Intelligent Deceleration Fuel Shut-Off). Six-speed machines couldn't have MDS. Challenger R/T, with the standard 18in rims, had 3.06 final drive ratio as an automatic, and 3.73 ratio with the six-speed stick. 20in rimmed six-speed cars came with a tighter 3.92 rear end. The adventurous could specify an R/T 'Track Pak' option group with Tremec six-speed, lsd, and self leveling rear shocks.

Made at Brampton Assembly in Canada, deposits were taken on December 3, 2007. Those initial 2008 MY cars were all SRT8s, numbering 7119. The 6400 cars allocated to America were all pre-sold, many of them much higher than the $40,095 list price. The power team involved was a Hemi 6.1 V8 and five-speed Autostick. The new coupe made a simultaneous debut at Chicago and Philly Auto Shows. SRT8's standard power team commenced with the 372 horse Hemi 5.7 V8, with 398lb/ft for the five-speed auto.

Let's not say 'Paint & Tape,' but 2009 saw a Rallye Package for Challenger SE with dual hood and trunk body stripes, chromed gas filler door, rear spoiler, 18in alloys, and Micro Carbon interior accents. Special Challenger packages were legion. The Challenger R/T Classic's retro theme involved Heritage colors like Toxic Orange, TorRed, and infamous Plum Crazy Purple. Factory strikes limited the Spring Special SRT8, painted B5 Blue, to a mite over 250 units. Also running in 2009 was a Drag Race Package, for NHRA competition.

This drag pack was based on a lightened Challenger SRT8. 1000lb were saved via composite polycarbonate material and lightweight parts. The engine was repositioned, and one half-inch was taken off the regular 116in wheelbase. The front cradle had a bolt-in crossmember, and solid engine mounts.

To commemorate the 50th anniversary of Grand Spaulding Dodge, Mr Norm's 50th Anniversary GSS Challengers were released. (Courtesy Larry Weiner, mrnorms.com)

The modern Challenger is based on a truncated LX platform. In LC chassis form, the current Challenger evokes the nostalgia of the 1970-74 E-body original. (Courtesy Larry Weiner, mrnorms.com)

There was a 50 unit homologation requirement. Engine choices were Hemi 5.7, Hemi 6.1 and 5.9-liter Magnum Wedge. The live axled racer had manual or automatic gearbox choices. Don Garlits bought the first car ... and raced it! The prototype SEMA cars, and production run, were finished by MPR Racing of Michigan. The base cars were supplied to MPR by Dodge. Completed cars could do high-nines at 135mph.

The limited 606 unit run of Mopar '10 Challenger R/Ts were painted Pearl Metallic Black, with a choice of three accent color stripes: blue, red, or silver. Black R/T Classic 20in rims joined a Hurst Pistol-Grip shifter, custom badging, Mopar cold air intake (worth 10bhp), and Katzkin aftermarket interior. The base car was Brampton Assembly built, and then sent to Mopar Upfit Center in Windsor, Ontario. America was allocated 500 cars and Canada got 100 units. Automatic versions numbered 320, and manuals 180. There were 255 blue accent cars, 115 red, and 130 with silver decoration.

Challenger received 2011 upgrades in line with Chrysler 300 and Dodge Charger. Base Pentastar 3.6 V6 was 305 horses strong. The 2011 Challenger Rallye package brought premium leather, and unique Foose design Rallye rims. The R/T got a new bottom grille cutout and reworked suspension. The latest SRT8 received the new Hemi 392 (6.4-liter), plus a new air dam influenced by the '70 Challenger R/T. This front spoiler functionally improved downforce. The Hemi 392 was tuned for torque; high-output OHV V8s usually sacrifice torque, due to excessive valve overlap. The Challenger R/T Hemi 5.7 was good for 0-60mph in five seconds, the SRT8 took 4.5 seconds.

In 2012, the SE variant was retitled SXT. For value seekers, the 2013 Rallye Redline V6 package brought exterior accents, black chrome 20in rims with Redline accents, sports suspension, better brakes and 3.06 final drive ratio. Radar Red Nappa leather interior and EVTS were options. EVTS, or Electronic Vehicle Tracking System, was a GPS-based stolen vehicle recovery system. 2015 saw the five-speed auto upgraded to a new ZF 8HP box, the Hemi 392 moved up to 485 horse and 475lb/ft. The new split grille, reminded fans of the '71 Challengers; ditto the new taillights.

Quad LED Halo Ring headlights, LED taillights, and functional Hemi hood intake were further 2015 improvements. A TFT (Thin Film Transistor) display now handled screen interfacing. SRT8 had six-piston front calipers and four-piston rear callipers, gripping two-piece 15.4in slotted and vented disks of the Brembo braking system. The new 707bhp 2015 Challenger Hellcat had its inner left driving light replaced by an air intake, and two colored key fobs: black for 500 ponies, red for the full 707 horses. Hellcat could do high-tens on street legal drag rubber; 0-60mph in 3.6 seconds on normal tires, and 60mph to rest in 109ft.

Hellcat could pull 0.94g on the skidpad and exceed 200mph! Even so, third parties saw room for improvement. The SMS 570 and supercharged 570X claimed 700 horse. Mr Norm's

The Mr Norm's Challenger-based GSS 'Cuda Racer Edition recalls famous E-body Plymouths of yore. Note the '71 D-segment tribute grille! (Courtesy Cliff Gromer, mrnorms.com)

Challengers were in the 635 to 900bhp range. The supercharged SpeedFactory SF600R was on 600 horse, and the Richard Petty Signature Series claimed 610bhp. For style, try Legacy by Petty Convertible Challenger, with custom front fascia, and NASCAR embellishments. In its first full year of 2009, Challenger managed 34,854 sales; in 2015, it reached a 66,365 total.

A demonic comeback – Challenger SRT Demon

Autoweek was surprised Dodge chose the Demon moniker for its new limited production 2018 MY Supercar. The original A-body Demon ended in 1972 under pressure from Christian conservatives. More than this, the old Demon logo was a cute cartoon devil, with stylized pitchfork. The new Demon logo is a scarier rendition, with a resemblance to Lenin. *Autoweek* remarked on the graphic nature of the new logo, too. However, the new car certainly lives up to its name, with hellacious levels of performance, which would have been unbelievable in 1972.

Building upon the Hellcat's basis, the Demon is a lighter, wide-body Challenger, with 2.7-liter blower for its 6.2-liter Hemi V8. It's rated at 808 horse on 91 octane juice, with 770lb/ft using 100 octane fuel. At 4280lb, it undercuts Hellcat by 208lb. Utilizing Nitto NT 05Rs sized 315/40R18, the Demon is the first production car with drag radial road tires; and there were other firsts. With a device putting the automatic in first and reverse for launches, the Demon is the first production car with an automatic pre-loading trans-brake.

It set new records for 0-60mph (2.3 seconds) and quarter-mile (9.65 seconds), concerning non-electric road cars. At 1.8g, it's also the hardest launching car to date, and first ever to do a wheelie! The NHRA banned it, because a car this fast needs a NHRA certified rollcage. For an extra buck a time, Dodge will put in a passenger seat and rear bench. The standard Demon has a rear rollcage, and

The 2018 Dodge Challenger Demon is so fast, the NHRA banned it for lack of an approved rollcage. The NHRA has always been rule-rigorous with Mopars! (Courtesy Dodge, dodge.com)

four-point driver's seat harness – and it needs 'em!

The Demon made a 2017 New York Auto Show debut, with planned 3300 production run of sub $100k units for 2018 model year. Production was planned for summer 2017. Yet another first, Demon's a/c system routes air to the intercooler for an even denser combustion chamber air supply ... cool!

Challenging the track

From 2010, Challenger R/T entered NASCAR'S Nationwide Series Competition. Factory involvement ended in 2012, with privateers continuing thereafter. JJ Yeleg ended 2014 with many top 20 finishes, and some top 10s. In 2015 and 2016, Mike Harmon and Derek White raced #74 and #40 Challengers, respectively, in the renamed Xfinity series.

Moving to the Trans Am TA2 series, Cameron Lawrence and Joe Stevens campaigned their Miller Racing Challenger. This racer had SRT and Mopar support, and used a Howe road racing tube chassis, and fiberglass body, plus 500 horse Hemi 392. Lawrence won four of the twelve 2015 races, coming third in the series. Joe Stevens came sixth. In March 2017, Challenger returned to the fastest class of Trans Am racing, for the first time since the '70s. Jeff Hinkle drove for the American V8 Road Racing Team. John Debenedictis served as crew chief.

Mr Norm's Grand Spaulding Dodge *youtu.be/ro9Q0w66lc0*

A seminal figure in Dodge Dart's leap from economical senior compact to muscle car was Norman Kraus. In the fall of 1962, 28-year-old Kraus opened Mr Norm's Grand Spaulding Dodge, located at the intersection of West Grand Avenue and North Spaulding Avenue, in the Chicago Humboldt Park area. Industry types said a dealer that specialized in high-performance youth models wouldn't fly. However, Norman Kraus accurately forecast the winds of change in the windy city. His famous dealership had its genesis in his family's gas station of the late 1940s, which had a used car lot. By the late '50s, this ancillary business started to focus on stick shift and performance cars. Kraus could see that not all buyers were family-type guys looking for a lead sled sedan, or vacation wagon.

Norman Kraus predicted correctly, and his Grand Spaulding Dodge (GSD) business saw sales double every year between 1963 and 1971. AM radio rock channel 'WLS Chicago' and word of mouth helped spread the good high-performance news. The 'Mr Norm' moniker arose from an initial newspaper ad space limitation. GSD was one of the first new car dealers to install a chassis dyno in the service department. When customers wanted a performance improvement, this apparatus proved they were getting their money's worth. In 1966, business was doing so well, a second Clayton dynometer was installed. 1967 saw the launch of a seminal model, which prompted Dodge to follow in hot pursuit! It was the Dart GSS (Grand Spaulding Special). Mr Norm gave the story on a TV show about Mopar history: with buyer demands for a hot Dart, he asked Dodge if he could have a big block car. The engineers said the big block wouldn't fit, so no deal. GSD service department folk then put a 383 crate motor into an ordered '67 Dart, and presented it to Mr Norm for a test drive. Mr Norm told them it was beautiful, and asked what they had done? The technicians then explained how they snaked the 383 into 'lil 'ol A-body, using relatively simple mods: the service department had repositioned the engine mounts, shaved the A-frames, and put heat shields over the exhaust manifolds. Exhaust manifold to engine compartment clearance had been the chief concern of Chrysler engineers.

The Dart GSS 383 was then shown to Chrysler executives. They were mighty impressed, and gave the green light to the factory-built 1968½ Dodge Dart GTS, with optional 383. Dodge's lack of a pony car and big block in the compact class was solved. It could now take on Camaro SS 396. Beyond this, GSD tech boys used their Clayton dyno to heat the 383 upwards of the 300-horse that was manufacturer claimed. Always the man in front, Mr Norm then ordered 50 '68 Darts, and a like number of 440 crate motors. As before, the GSS 440 inspired the 1969 'M' code GTS 440.

Mr Norm's first 1968 Dart GSS 440 was the prototype for the factory 'M' Code Dart GTS 440. (Courtesy andy440.com)

Yeah, dealers have demonstrators, but how many dealers have a red, worked Hemi Coronet R/T convertible? Mr Norm did, and if racing improves the breed, it can also promote it: 'I want what he's racing!' In 1964, GSD went racing, with standard wheelbase Max Wedge and Hemi powered cars. The trouble was, this put GSD up against its customers, and you don't want to beat them; so, in 1965, GSD went altered wheelbase, with a blown Hemi 426 motorvated Coronet. Gary Dyer set records across the land, often beating the factory Ramchargers, and did an 8.63 second pass at Lions Dragway in Long Beach, California.

In 1969, Mr Norm's Dodge Super Charger won the Coca-Cola 'Cavalcade of Stars' Championship. GSD was also known for its 1971 Super Stock Demon, but Mr Norm could always see the writing on the wall. Part of GSD's success was its diversified business interests, recognising new trends. The NHRA was putting the bite on the all conquering Mopar Marvels, and the muscle car sector was generally past its sales peak. To combat rising insurance premiums, and follow into downsized muscle, GSD had its '71 Demon 340 six pack and '72 blown Demon 340 SS.

Always way out in front, Mr Norm recognised that the custom van and truck market was on the rise; GSD was into that as early as 1970. In 1973, GSD had a formal department for van conversions. The previous year, Mr Norm's Kawasaki was instituted. Kawasakis were green in those days; so too were the dollars GSD was earning. Forward thinking saw GSD reach 1972's No 3 top selling Dodge

Neither Mr Norm nor Dodge could provide power steering or power brakes, due to A-body space constraints. With no room for a booster, front disks were off the menu. (Courtesy andy440.com)

dealership in America. In 1973, it reached No 2, and in 1974, Mr Norm's Grand Spaulding Dodge was the No 1 Dodge dealer in the universe!

Chrysler Corp always majored in fleet sales – maybe a little too much – and GSD was strong here too. By 1975, its fleet department was so big, it supplied all Chicago Police Dept cars, Illinois Secretary of State, Department of Transportation, and more. Special order Dodge Magnum 440 V8s were a specialty. GSD dyno tuned cop sedans, and their work was enjoyed by Jake and Elwood in *The Blues Brothers* movie. Sadly, with the whole high-performance thing winding down, Mr Norm sold his interest in GSD in 1977. The dealership continued as a normal car dealer into the late '80s, when it was finally closed. The building is still standing at the same address.

Mr Norm returned in the 21st century. In 2006, a range of diecast models from Supercar Collectables was released, followed by a limited run of full-size 1968 GSS Hemi Darts. The Florida-made, serial numbered run of 40 cars, was represented at the 2006 SEMA show's Mr Gasket company exhibit. In 2008, a run of 100 Dodge Ram Super Trucks, called 'Mr Norm's Super Charged 500' were done by Larry Weiner. Subsequently, Mr Norm performance packages were offered for modern Mopars. Weiner ran this limited edition vehicle and parts program for Norman Kraus.

Larry Weiner, editor of *Pentastar Power* magazine, also revived the Mr Norm's Sport Club. 2013 marked the 50th anniversary of Grand Spaulding Dodge. To commemorate this milestone there was a Mr Norm's 50th Anniversary National Tour. This did the rounds of convention centers. Car-wise, there were special Mr Norm's 2013 Dodge Challengers and Chargers. The Mr Norm's speed shop phenomenon – something insurance premiums, two fuel crises, and the EPA couldn't stop!

The iconic Bumble Bee stripe. It took Mr Norm to get the A-body beyond the 273 V8, and deliver a real sting in the Mopar tail! (Courtesy andy440.com)

Without power steering or power brakes, Mr Norm's Dart GSS 440 was a muscle car in more ways than one. Grand Spaulding Dodge competed against customer cars in standard wheelbase drag racing. (Courtesy andy440.com)

Mr Norm's Norman Kraus was a speed shop pioneer, being one of the first to introduce a dyno at a new car dealership. Grand Spaulding Dodge could prove your car had more muscle! (Courtesy andy440.com)

Mopar Missiles & Hemi Colt

By the '70s, Mopar drag racing dominance was coming under attack. However, the pressure wasn't so much from GM and Ford as it was from non-racing circumstances; basically, NHRA rulebook changes to limit Mopar winning ways, and economic events that made racing dollars scarce. The Mopar Missile program involved a team of engineers and racing professionals campaigning a Chrysler Corp sponsored 1972 Duster, updated to '73 MY look. This Mopar Missile won 11 out of 14 major 1973 NHRA events.

This was the third car in a 'Missile Trilogy.' There was originally a '70 Challenger, updated to '71 MY style, that raced as the 'Motown Missile.' This was followed by a '72 Barracuda. The name change to Mopar Missile came due to a clash with Motown Records. Ironic, given Chrysler Corp and the record label would hook up decades later, concerning special edition LX platform Chrysler 300s. Tom Hoover, associated with the Ramchargers, and getting Hemi 426 V8 into racing, was project leader on Mopar Missile.

Racing, in those times, wasn't exactly unrestricted. According to the recollections of team member Stewart Pomeroy, recounted by AllPar's Gene Yetter, the cars ran limited. Engine size was $400in^3$, no power adders, weight was around 2700lb and times were high-eights at 155mph. Interesting touches were computer data analysis and budget aero work. The former involved a trunk-mounted electronic data analysis unit. This was a recorder using tape or wire, that monitored eight parameters.

With the NASA space program winding down by 1971/72, some personnel took the offer to come over to Chrysler, and help with drag racing. The trunk computer was the upshot. To avoid the cost of wind tunnel testing, cotton strands were placed over the Duster's body, and the car was photographed at

The drag racing Missile car trilogy ran from Dodge Challenger to Plymouth Hemi 'Cuda, and Duster. This picture was taken by Gene Yetter at the 2012 Carlisle Show. (Courtesy hobbyexpress.com)

speed. Pictures were studied to see how closely the strands were pressed against the body. Less turbulence was preferred. The fuel crisis, economic downturn, and Chrysler's general business problems caused the corporation to quit Pro Stock in 1974. The Mopar Missile then went to racer Don Carlton, and subsequently spent 18 years in Don Garlits' museum.

With the NHRA playing with the rulebook to foil Mopar/Hemi V8 success, something had to be done. The NHRA had allowed small cars in Pro Stock, and rule change weight breaks gave them an advantage. This was compounded by Vegas and Pintos using longer wheelbases.

To combat these small block rivals, Chrysler Corp turned to car development using its LA small block V8. The 'Wire Car' was Chrysler's first gen small block Pro Stock car. The last of the test cars at the factory's Pro Stock racing program. The Yellow 340 car and Hemi Colts were along similar lines, as weight break rule fighters. Chrysler even got Ford man Bob Glidden to do an LA-engined Plymouth Arrow subcompact.

Concerning passenger road cars, Chrysler thought subcompacts were a fad. While cars like AMC Gremlin and Ford Pinto sold up a storm, Walter P relied on captive imports. First off, the Plymouth Cricket, and then, from 1973, the Dodge Colt from its Mitsubishi link. Chrysler Corp wanted fair weight break rules, so it could use destroked Hemis in A-body. For example, in September 1972 at the York US30, Ronnie Sox set the quickest ever Pro Stock match race times in his Duster: 9.17, 9.27, 9.16 seconds.

If the NHRA didn't play fair, Ted Spehar's Hemi Colt would be an avenue for victory. Chrysler's last resort was to use 'em in AHRA, IHRA, and match race, like the early funny car days. The NHRA had qualms over the imported car nature of a destroked Hemi Colt.

The Dodge D5 Hemi Colt was a Chrysler Corp last resort to overcome unfair NHRA rule changes designed to undermine Mopar Hemi Pro Stock dominance. (Courtesy Bill Crittenden, carsandracingstuff.com)

Using destroked Hemis, Chrysler Corp hoped to stay competitive against domestic small block subcompacts, which had been given favorable weight breaks. (Courtesy Bill Crittenden, carsandracingstuff.com)

So did some readers of *Super Stock & Drag Illustrated*. The magazine's editors pulled a fast one with the April 1971 issue, with the headline 'Dodge's New OHC Hemi.' A cover like that really boosted circulation. Trouble was, when readers enquired within they found a four-cylinder captive import! The editors did print an apology in the next issue.

Still, needs must when the Demon doesn't drive. In mid-1973, Mr Norm got Romeo Palimedes to build a Hemi Colt. Colts were being built for Pro Stock, and the Modified Compact Class, involving hotted up inline fours and V8s. However, Chrysler Corp was still getting by on A-body, so no Colt factory support in 1974. Chrysler boycotted NHRA Pro Stock in 1974, due to the governing body's rule changes. The factory offered true wedges in 105in wheelbases or greater, so avoiding Hemi Colt, but there were no takers. However, non-factory Hemi Colts were already running match races.

Come August 1974, at the PRO National Challenge in Long Island, Ronnie Sox raced in the Don Hardy-built Ronnie Lyles Colt. Ditto, Bobby Yowell in the Billy The Kid/Stepp Colt. Sadly, these Hemi Colts weren't competitive against lighter rivals. In 1975, the Colts were doing well in the Sportsman divisions, especially altered classes of Competition Eliminator. Mr Four-speed did an 8.60 second pass at 159mph in a match race Colt. Don Carlton's gold Rod Shop Colt at Pomona in 1976, in B/A class did an 8.4 second quarter-mile at over 160mph. That's one Colt the EPA never tested!

The Hemi Colts gained a reputation for being very squirrely concerning handling. Tragically, Don Carlton died testing a Colt for Chrysler at Michigan in July 1977. Rule changes where 92in would be the minimum wheelbase, and the greater stability of small blocked Plymouth Arrow, led to Ronnie Sox, Roy Hill, and others switching from Colt to Arrow. These cars offered favorable weight breaks, with the LA 340/360 V8s. However, Ronnie Sox and Clyde Hodges, with some money from Buddy Martin, gave Hemi Colt another try early in 1978. This was the late Don Carlton's gold colored Indy '76 winner, and had the Colt body swapped for the Dodge captive import Challenger body. Unfortunately, this Colt based car wasn't competitive.

Then it was Arrow all the way; Bob Glidden drove one, and dominated to win the NHRA Pro Stock title in 1979. No Mopar money caused Glidden to return to the Ford fold in 1980. Dandy Dick Landy got in on the LA block action, before retiring at the close of the 1981 season. Ronnie Sox moved to the Dean Thompson Mustang, and won the 1981 IHRA title. Racers and fans struggled to get a handle on Pro Stock Hemi Colts back in the '70s, but they are popular today. How would you describe a little Japanese captive import with a big block Hemi V8? The Japanese have an

Several private operators used Hemi Colts in the 1973-77 era, including Ronnie Sox. This Hemi Colt was tuned by Sox & Martin chief mechanic Jake King. (Courtesy Bill Crittenden, carsandracingstuff.com)

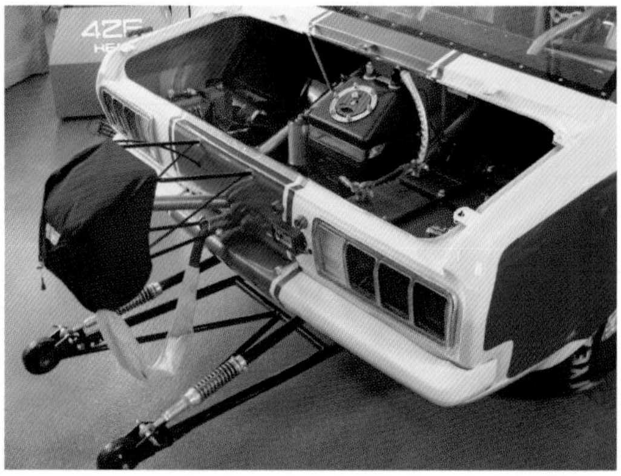

Don Carlton got down to an 8.4 second pass in the Rod Shop Colt, but the little monsters could be tricky handlers. (Courtesy Bill Crittenden, carsandracingstuff.com)

expression for that: 'Chou sugoi!' It means 'far out!'

Mopar Missile number four was the Dodge D5 Hemi Colt, but number five was Aspen- based. Don Carlton owned the Mopar Missile name, and he teamed up with long-time collaborator Clyde Hodges on this special Dodge Aspen. Missile No 5 was built in their Lenoir, NC shop in 1977. It would be their last project. The car was never successfully raced, and lay in storage for much of its life. Subsequent owner Mike Ricketts restored the Aspen back to its original state. Ricketts received permission from Don Carlton's widow, allowing him to rename the car Macomb Missile. Macomb was Ricketts' neck of the woods.

The fifth Missile car was the last project Don Carlton and Clyde Hodges created. The 1977 Dodge Aspen was renamed Macomb Missile by the car's restorer, Mike Ricketts. (Courtesy Gene Yetter)

Mopar Motored V8 Exotics

In the 1983 movie *Cannonball Run II*, the Lamborghini girls hitch a ride from a gallant knight driving a Mercedes 300SL Gullwing. They notice it doesn't sound stock. He tells them the stock inline six has been replaced with a 383 V8, and now it really moves! That wasn't the first time someone swapped in good ol' American OHV V8 iron, into a European chassis – small block Chevy V8s in Ferraris, BMW 507s etc; 'Take that Porsche and Chev it!' was a slogan from one conversion company. In the 1990s, Converse Engineering did Ford 5.0 V8 Mustang conversions for Volvo 240 and 940. Carroll Shelby had a similar idea with the AC Ace and Cobra.

Sometimes the original Euro motor was worn out, expensive to replace, or a found exotic was missing a motor. Oftentimes, owners were just looking for easy, affordable power. Back in the 1950s and 1960s, exotic European and British brands turned to American V8s. Reliable, high horsepower, such V8s went great with a Euro suspension and braking system. It was a great way for small companies to get on terms with the bigger boys, like Ferrari and Porsche, which rolled their own. Frequently the quality and power of Mopar muscle, made Chrysler Corp their go-to company.

The hot Humber

Chrysler formally took control of the British Rootes Group in 1967, as part of Chrysler Europe. In the years leading up to this, Chrysler's stake and influence in the British concern was rising. The Humber Super Snipes were one of Britain's leading director/management sedans, with the Imperial as pinnacle model. Sir Reginald Rootes had a custom Imperial created for his use in 1966/67. Just behind the front fender, and under the Imperial chrome script, lay a V8 plaque badge. It signified Chrysler's 180 horse LA small block 273 V8 underhood. Chrysler UK was replacing the luxury Humbers with V8 Valiants from Chrysler Australia. Sir Reginald switched to one of those Aussie cars, post Imperial. Such Valiants had the circumstance, but lacked the Humber's pomp and presence.

Bristol

Located in Filton, near Bristol, this British company had an aviation background.

Its early postwar start with car making involved a gentleman's racer with BMW 328 styling, engine, and chassis influence. However, it soon became associated with owner Anthony Crook, and Chrysler V8 power. Boss and owner Crook was a WWII fighter pilot, and a racer. He said the company came across Mopar muscle when looking for a new automatic transmission.

The company's first V8 car was the 407. At the start of the '60s, Bristol had been working on a new 3.8-liter OHC alloy six for the 407, and needed an autobox. The Filton firm had intended to do its own automatic but, running late, rang up Chrysler for a TorqueFlite. The gearbox came with a Chrysler V8 attached. The motor worked so well the company decided to use both the TorqueFlite and Chrysler V8! It developed the 407 around the Mopar V8, but improved said V8's durability.

The stock V8 was good for some 50,000 miles in high-speed 100mph European driving, but Bristol was used to 150,000 miles, so it redesigned the oil pan with well placed baffles. Chrysler was pleased with the result. In Crook's words: "They were delighted."[11] In the age of the automotive character, Crook was as distinctive as his cars. A real adventurer: in the '70s he would fly his plane for testing of Bristols with run-flat tires, then fly back to London. Enzo Ferrari used a pen with lavender ink for all correspondence; Crook used green ink, and wrote polite letters to magazine editors when he felt journals had assessed Bristols incorrectly. Being a small company with a limited international dealer network, he liked the Chrysler powertrain for its efficiency, durability, and power. It was also fixable by any competent mechanic.

Bristol later turned to Italy's Zagato for styling, and often used the 383 V8 and 360 from 1978. In the late '70s, Bristol was hit by pollution controlled V8s, and in 1977 used the humble 318 V8 as an entry level fuel crisis necessity. However, in the '80s the firm fought back with turbocharged 360 power! Between 1946 and 1994 it had made just 8004 handbuilt coupes, with independently sprung big separate chassis.

Monteverdi

Switzerland had a car maker called Monteverdi, which used Chrysler V8 power big time. Peter Monteverdi was a Swiss racer, and luxury car importer/dealer. He was the Swiss Ferrari agent until late 1965. Just as with Ferrucio Lamborghini's argument with Enzo led to the Lambo bull, disagreement with Enzo pushed Monteverdi to make his own Ferrari fighters. There was the 1967 375S, 2+2 375L and 400 horse 400SS. Over 1971-77 Monteverdi's 375/4 was a 150mph Maserati Quattroporte rival four-door. Frua did the early styling, but a disagreement saw a change to Trevor Fiore, of Fissore, after 1968.

Monteverdi liked a good strong chassis, feeling this to be more important than lightness. His cars were available with Chrysler 440 and Hemi 426 V8 punch, including the four-door limo. However, his most famous model was a mid-engined sportscar that never officially went on sale: the Monteverdi Hai – 'Hai' was German for shark.

The Hai had a heavy tubular chassis, with wishbones, coils, and swaybar at the front. At the rear was a de Dion tube, Watts linkage, and upper and lower trailing arms. Konis and disk brakes were at each corner. Monteverdi always asked for the most powerful motors; in this case, a solid lifter Hemi 426 with 450bhp at 5000rpm and 490lb/ft at four grand. The engine sat partly in the cabin, with leather-trimmed insulating cover. Drivers worked a five-speed transaxle, as Pirelli GR70-15 shod Borrani wires whisked the Hai to 176mph, after hitting sixty in 4.7 seconds. This was *Automobile Quarterly*'s experience.

The beautiful Hai had Monteverdi himself claiming styling credit, although Trevor Fiore is a more likely main

The desirable wares of Swiss automaker Monteverdi. In the factory workshop is a red 375L 2+2, yellow 375/4 limousine and Hemi V8 motor. (Courtesy monteverdi.ch/en_museum.html)

contributor. A short wheelbase, Purple Smoke-painted Hai with white leather interior made a March 1970 Geneva Show debut. It toured for 18 months, repainted and trimmed a few times. Proposed price was 27 grand, but some say the car was only ever a come-on to generate publicity for the marque. Either way, it handled and accelerated as well as any exotic. In 1973, a 50mm longer, bright red 450 GTS, with 440 power and lower rocker panel Monteverdi black decal callout, did the show circuit. As of 2015, the original Purple Smoke car, resided with a private collector. The other three Hais are in the Monteverdi museum.

Less exotic was the rebodied F-body called Monteverdi Sierra, plus two-door convertible conversion of the M-body Dodge Diplomat. A one-off Plymouth wagon based car was never sold. Monteverdi did sell many Safaris and Saharas; these V8 SUVs were rebodied International Scouts, with Fissore styling help. Using standard Scout 345in^3 V8, but always Chrysler A727 TorqueFlite and sometimes 440 V8. Around 1000 of these SUVs were built. Eventually Monteverdi moved to less commercially successful rebodied Range Rovers. As a car maker, the Swiss company ended in 1984. The factory became a museum the following year.

Jensen

'Jensens for gentlemen –' that was the slogan of a paternalistic company. The founders were addressed and spoken of by employees as 'Mr Alan' and 'Mr Richard'. Based in West Bromwich, England, Jensen was a larger concern than Bristol, but still very small. In 1969, out of Chrysler Corp's 1,805,846 unit total, just 400 Jensens were made, with

Peter Monteverdi believed in a stout chassis for the Hai 450SS. It made Miura seem a mite wobbly!
(Courtesy monteverdi.ch/en_museum.html)

The amazing Purple Smoke-painted Hai 450SS, on debut at Geneva in 1970. Planned price was $27,000! (Courtesy monteverdi.ch/en_museum.html)

Chrysler V8s and TorqueFlite/four-speed drivetrains.

The company met up with Mopar when developing its Jensen 541-S. Jensen planned to use an Austin 4-liter straight six and Rolls Royce GM Hydramatic as a drivetrain. European luxury customers were increasingly demanding automatic transmission – cheaper than a chauffeur! The Austin motor didn't work well with automatic, so Jensen went to the 305 horse 383 and TorqueFlite for its new fiberglass bodied C-V8 coupe. The car was the transport of John Mannering, the adventuresome antique dealer played by American actor Steve Forrest in *The Baron*. *The Baron* was a 1965-66 TV show, made by ITC.

After the C-V8, Jensen moved on to the steel-bodied 'Touring of Italy'-styled Interceptor. A typical Interceptor spec going into 1970 had the 383 V8 on 10:1 CR with one 4bbl Carter, rack-and-pinion steering, four-wheel 11.25 in Dunlop disk brakes, rear leaf springs, TorqueFlite and PowerLok lsd 3.07 rear axle. At this hour, Bristol was using Canadian spec V8s to get around US smog law.

CAR magazine's Mike Twite offered insightful comments in the September 1969 issue. He said he saw 130mph so many times on the speedo, it started to feel slow! The motor never overheated, used hardly any oil, always started first time, and was almost silent. However, on his race reporting jaunt to Belgium, the 383 V8 rated at 325bhp at 4600rpm had one problem: "... a thirst as bad as that of an Irish labourer on a Saturday night."

That thirst would prove to be a problem with the fuel crisis. Up to that point, the Interceptor had done good

Independent testing of the Hai 450SS produced a sub-five second zero-to-sixty time, and almost 180mph. (Courtesy monteverdi.ch/en_museum.html)

business, and had star power as well as Mopar power. Tony Curtis had visited the factory, and owned an Interceptor, when he was doing the TV show *'The Persuaders!'* in 1971-72. By the '70s, Jensen was owned by Kjell Qvale and the Healey family. Shared interests cooked up the Lotus 16 valve, I4-powered Jensen-Healey sportscar. It was even featured in an episode of *Here's Lucy* in the early '70s. Here it was a game show prize, with 8-Track player, for Lucie Arnaz.

Jensen's other star project, was the AWD Interceptor FF. This had Ferguson Formula all wheel drive, and Dunlop Maxaret ABS brakes. The car was expensive, and a slow seller. Just 15 MkIIIs were sold, after a debut at the 1971 Earl's Court Motor Show. It's replacement was the Jensen Interceptor MkIII SP. This car had a special six pack 440 supplied by Chrysler in 1972 and 1973. The triple 2bbl Holleys and fine touches like forged crank, rods, pistons, and big cam, spelt 330 horse net.

The hot cam with longer duration, was needed to generate enough vacuum to bring the outer carbs online. The Holleys

Driver and passenger sat between the front part of the Hemi 426 V8, which lived under a leather-trimmed cover casing. So, in spite of the a/c, the interior felt cosy. (Courtesy monteverdi.ch/en_museum.html)

The original Geneva show car was repainted and trimmed, touring the show circuit for 18 months. (Courtesy monteverdi.ch/en_museum.html)

Monteverdi did rebodied Sahara and Safari SUVs, based on the International Scout. These successful SUVs sometimes had 440 V8/TorqueFlite A727 power teams! (Courtesy navistar.com)

need adjustment every 1000 miles via an air screw. Each carb has an independent float level setting. This makes them more prone to flooding, and a possible fire risk. Unfortunately the SP's sales fire was out, and the Jensen-Healey was unreliable and bland looking. The Interceptor V8 was hit by the fuel crisis and ensuing recession. Mr Qvale liquidated Jensen in 1976.

Monica

French railway carriage tycoon Jean Tastevin wanted to make a car; a modern Facel Vega. In the early '70s he almost did. Swoopy Jag XJ6 profile, Corvette C3 front fascia, tubular frame chassis and de Dion suspension made it look like a dream car. Sadly, the fuel crisis turned the dream into a nightmare, and curtailed production plans for the four-door sports sedan.

Facel Vega

Probably the most expensive new car the 383 V8 ever got used in, the French Facel Vega was what one owned when you had everything else. It was 383 and TorqueFlite, or four-speed, but said four-speed was a French box. The heavy duty Pont-a-Mousson four-speed came from a truck, and was unlikely to be specified by wealthy clientele like Ringo Starr or Francois Truffaut. Manual power teams were tuned to 390bhp at 5400rpm, automatics were on 355bhp at 4800rpm. Top speed was nearly 150mph, with 0-100mph in 17 seconds. However, it was the luxury and exclusivity that counted. Handbuilt light body panels, leather and wood interiors, all craftsmen created. The final 1962-64 Facel Vega II was all that and more. It even pulled off the stacked-headlamp Mercedes-Benz look ... perhaps better than Stuttgart.

Chrysler Overseas

In the opening episode of the third season of *Get Smart*, there was a little product placement by Walter P. The bad guys turn up in a French Simca to ambush Max in his red Sunbeam Tiger. Then, they all pile into a new 1967 Plymouth Barracuda to comically drive a few feet. The Barracuda was red, to help one forget that blue Karmann Ghia Max was always pulling up to the magic phone booth in.

To power one's Chrysler Courier 229 Hydro-Vee hull speedboat, Chrysler Marine offered an 80 horse Simca inline four cylinder job as starter powerplant. At the top, a 290bhp Hemi 426 looked like overkill! Well, with the formation of Chrysler Europe in 1967, Walter P was calling on all parties. In 1967, the Simca GLS was a rear engined captive import, with 60in^3 engine to power a car 149in long and 1710lb in weight. No need for shade tree dealers and mechanics, like '50s imported small cars. Simca carried Chrysler's 5 year/50,000 mile power train warranty stateside, and could draw on an 800 outlet dealer network. Even so, it was a quirky small car. VW offered a larger car, with a big service network.

That was the French side of Chrysler Europe. The British part was represented by the Rootes Group of brothers. Billy Rootes – Lord Rootes, eventually – had made American trips, and adapted US styling trends to Rootes cars. This soon included observation of '50s Plymouths and Chryslers penned by Virgil Exner. The 'Million Dollar Look' became the 'Million Pound Note,' as it were. Soon the link between Rootes and Chrysler turned from one of influence to necessity.

Rootes' plan to make a sophisticated Mini rival in an area of high unemployment and cheap labor – Linwood, Scotland – didn't work out. The Hillman Imp was late to market, it had mechanical gremlins and the workforce wasn't used to making automobiles. The Imp got an early reputation for unreliability. By the mid '60s, Rootes was struggling against Ford UK, Vauxhall (GM), and local giant BMC (Austin & Morris). Chrysler wanted into Europe, and Rootes was folded into Chrysler Europe.

Chrysler Europe's new big car project, 929XC saw a three way submission proposal. One came from Simca/Bertone, one from Rootes, and a third from Chrysler's International Studio (USA). Eventually the British proposal won, and was built in France as the Chrysler 180. The 180, and also new 1970 Hillman Avenger (Plymouth Cricket), had that late '60s Chrysler Corp small window, muscle car C-pillar look; plus, don't forget the earlier Sunbeam Rapier. The Rapier traveled to America with 100 horse 2bbl I4, disk/drum brakes, walnut dash trim, and reclining seats. It also had a coincidental visual resemblance to the 1964-66 Barracuda.

Then there was Spain; the Barreiros brothers had been making a version of the Dart from 1965 at their Villaverde, Madrid factory. It was called the 3700: a large, powerful Slant-6 executive machine. Using a CKD operation to avoid import duty, and local engineering and styling changes versus the US Dart. The 3700 was Spain's top selling car between 1965 to 1977. 17,589 were made from 1965 to 1970. However, the operation only made a profit one year, and the Barreiros brothers sold their operation to Chrysler Corp in 1966. This made it the Spanish arm of Chrysler Europe, or Chrysler Espana S.A.

Down under, Chrysler Australia had been an import-duty-beating CKD operation. Like Ford Australia, Chrysler was making US designs not really suited to local conditions. The 1954 Chrysler Royal was a V8, tailfinned, full-size sled, with rear Sports Deck trunklid. Too big, lavish, and complex. Not enough ground clearance and suspension toughness for unsealed roads, also. Like Ford, the

Peugeot-Citroen (PSA) bought Chrysler Europe for one buck in 1978. Remaining Simcas and Rootes group models were renamed Talbots. (Courtesy PSA, groupe-psa.com)

dawn of the '60s saw manufacture of a US compact that received more and more local engineering, distancing itself from its US counterpart. The car was the Chrysler Valiant, and it was a hit. Chrysler Australia did a Hemi Slant-6, put it into its scaled down Charger with four-speed and over 300 horses. Hemi 6 pak decals were abound, as was the catchphrase 'Hey Charger!'

Chrysler put their man in Europe. Harry E Chesebrough became Directeur General at Simca. Living in Paris, he was also responsible for Rootes. The late '60s appointment saw him try to make Simca less austere. By now, even European drivers wanted the high life. He also tried to harmonize and streamline Chrysler Europe: no more Sunbeam Imp, Bertone Simca Coupe, Sunbeam Alpine, or Tiger. Hopefully, Maxwell Smart would see the Sunbeam Fastback Coupe (Rapier) at its New York International Auto Show debut, and take 99's advice to trade in that Tiger.

Chrysler USA commenced late '60s design work on a subcompact ... then gave up on it. This could be because it was spending a lot on federal safety regs, smog controls, and money given to Chrysler UK to develop the Hillman Avenger. So why not use that as a captive import? If Walter P had a chance to read Steinbeck, he may have learnt what becomes of the plans of mice and men. Chrysler Europe and Australia got squeezed by GM and Ford's overseas subsidiaries, European car makers, and eventually Japanese companies. For the last, Chrysler Corp had only to look at its successful captive import relationship with Mitsubishi. Colt was a sure fire hit, with the economy and quality people wanted. Mitsubishi even bought out Chrysler Australia in 1982. In 1977, Peugeot-Citroen (PSA) had purchased Chrysler Spain. In 1978 it bought the rest of Chrysler Europe. The Plymouth Cricket had gone from Hillman Avenger, to the Chrysler Avenger and finished as the Talbot Avenger. It was the end for Chrysler Corp overseas.

Mopar's Marvelous Quartet

1962½ Dodge Dart Ramcharger 413 – Dodge street dynamite!

Base Dart 330 price: $2949
Ramcharger 413 package: $374.40
Engine: Cast Iron (head & block) OHV solid lifter 413in^3 V8, 11.0:1 CR, 2x Carter AFB 4bbl, 410bhp (gross) @ 5400rpm, Ram Induction aluminum eight-branch intake manifold - Premium Fuel.
Gearbox: HD Ramcharger Torque-Flite, Sure-Grip lsd with 3.91 rear axle.
Dimensions: 202in length, 76.5in width, 54.1in height, 116in wheelbase.
Weight: 3440lb
Chassis: Unibody
Front suspension: Independent non-parallel control arms, torsion bars, trailing torque strut, tube shocks.
Rear suspension: HD police six leaf-spring suspended live axle, tube shocks.
Wheels & tires: HD suspension included 14x5.5in rims & Goodyear 7.50-14s.
Brakes: Manual drums x4, (front) 11x2.5in (rear) 10x2.5in.
0-60mph: 5.8 seconds
Quarter-mile: 14.4 @ 101mph
Top speed: 119mph
Fuel economy: Not disclosed
Test data source: *Motor Trend* August 1962

By combining the new B-body midsize platform with the Ramcharger 413 V8 package, Dodge created an early muscle car. Also, a level of performance unsurpassed by affordable rides until the 21st century!

1969 440 'Cuda – the world's biggest engined pony!

Base Barracuda price: $2813 + 'Cuda Group (440 V8 Big Block, Torqueflite, HD Suspension, Wheels & Tires) $345 + Sure-Grip 3.55:1 lsd $42 + Tach $50 = $3931.	
Engine: Cast Iron (head & block) OHV hydraulic lifter 440in^3 V8, 10.1:1 CR, Carter AFB 4bbl, 375bhp (gross) @ 4600rpm, 480lb/ft @ 3200rpm – Premium Fuel.	
Gearbox: Torque-Flite A727, Sure-Grip lsd with 3.55 rear axle.	
Dimensions: 193in length, 70in width, 53in height, 108in wheelbase.	
Weight: 3405lb	
Weight distribution: 57.2 per cent (front)/42.8 per cent (rear)	
Chassis: Unibody	
Front suspension: Independent, short & long arms, torsion bars, swaybar (0.94in), tube shocks.	
Rear suspension: Six leaf-spring suspended live axle, tube shocks.	
Wheels & tires: HD suspension included 14x5.5in rims & Goodyear Wide Oval E70-14s.	
Brakes: Manual drums x4, (front) 10x2.25in (rear) 10x1.75in.	
0-60mph: 5.6 seconds	
Quarter-mile: 14.01 @ 103.81mph	
Top speed: 118mph	
Fuel economy: 9.8mpg	
Test data source: *CAR Life* June 1969	

The A-body 1969 'Cuda 440 was the largest engined pony car to that point in time. It out-accelerated later 440 and Hemi powered '70 Cudas in independent testing. However, it wasn't available with power steering, disk brakes ... or even power brakes! It was a high-performance car for the serious enthusiast.

1971 Dodge Demon 340 – devilishly good value!

Base Demon price: $2721 (includes 340 V8, Rallye Suspension, Tuff sports steering wheel, 150mph speedo & full instrumentation, all synchro three-speed floor shift, dual exhausts & Demon decals)	
Engine: Cast Iron (head & block) OHV hydraulic lifter 340in^3 V8, 10.5:1 CR, Carter TQ 4bbl, 275bhp (gross) @ 5000rpm, 340lb/ft @ 3200rpm - Premium Fuel.	
Gearbox: Torque-Flite A904, 3.55 rear axle.	
Dimensions: 192.5in length, 71.6in width, 52.6in height, 108in wheelbase.	
Weight: 3353lb	
Weight distribution: 56 per cent (front)/44 per cent (rear)	
Chassis: Unibody	
Front suspension: Independent, short & long arms, torsion bars, swaybar, tube shocks.	
Rear suspension: Leaf spring suspended live axle, swaybar, tube shocks.	
Wheels & tires: Standard Rallye suspension & E70-14s, 14x5.5in rims optional.	
Brakes: Manual drums x4, (front) 10x2.25in (rear) 10x1.75in.	
0-60mph: 7.8 seconds	
Quarter-mile: 14.56 @ 96mph	
Top speed: 127mph (estimated)	
Fuel economy: 14.3mpg city/16.7mpg hwy	
Test data source: *Road Test* April 1971	

The name courted controversy, however Demon's bang for buck was beyond debate. Having a Dodge kissing cousin for Plymouth Duster showed it was good to share. Demon and Duster 340s kept the fun levels high, and insurance premiums low. Dodge revived the Demon and Air Grabber concepts in 2017.

1975 Plymouth Duster – when bucks really counted

Base Duster price: $4115.53 (Includes: light package $31.05, tinted glass $43.70, dual mirrors $28.65, A/C $406.50, tape stripe $29.30, AM/FM radio $136.65, power steering $124.10, vinyl roof $75, power front disks $23.95)
Engine: Cast Iron (head & block) OHV hydraulic lifter 225in^3 I6, 8.4:1 CR, Holley 1945 1bbl, 90bhp (net) @ 3600rpm, 165lb/ft @ 1600rpm – Regular Fuel.
Gearbox: three-speed manual floor shift, 2.76 ratio rear axle.
Dimensions: 194.1in length, 71.8in width, 53.2in height, 108in wheelbase.
Weight distribution: 58 per cent (front)/42 per cent (rear)
Chassis: Unibody
Front suspension: Independent, short & long arms, torsion bars, tube shocks.
Rear suspension: Leaf-spring suspended live axle, tube shocks.
Wheels & tires: Goodyear Polysteel Radial ER78-14s on 14x5.5in rims.
Brakes: Power disk/drums, (front) 10.8in vented (rear) 10x1.75in.
0-60mph: 14.8 seconds
Quarter-mile: 19.47 @ 70.42mph
Top speed: 94mph
Fuel economy: 16.3mpg
Test data source: *Road Test* May 1975

By the mid '70s, the Slant-6 powered Duster was an inflation and fuel crisis fighter – a friendly and familiar bodystyle combined with reliability one could count on. It was America's Plymouth rock in difficult times.

Mopar Items Of Interest

AllPar – Dodge, Chrysler, Jeep, Ram etc. cars and trucks
www.allpar.com
The center of excellence for all Mopar matters. Includes an owners' forum.

Don Garlits Museum of Drag Racing
www.garlits.com
Don Garlits' Drag Racing museum. Showing the career and cars of the King of drag racing.

Mr Norms
www.mrnorms.com
Mr Norm's 21st century high performance Mopars, and Sport Club can be seen at this website.

Fujisawa Shonan Chrysler-Jeep
fujisawashonan.cj-dealer.jp/jeep/
Fujisawa Shonan Chrysler-Jeep. Your first stop for Chrysler and Jeep in Tokyo.

Slant-6 Forum
www.slantsix.org
An information resource, with forum, for Slant-6 and Chrysler A-body fans.

For E-Bodies Only
www.forebodiesonly.com
A forum just for E-bodies, the classic Plymouth Barracudas and Dodge Challengers of 1970-74.

Aspen and Volare
www.aspenandvolare.com
Eric Bremer's F-body forum and information resource site. Dedicated to the 1976-80 cars that evolved from the A-body, and preceded M-body.

Challenger Talk
www.challengertalk.com
A forum for the modern LC Challenger, with merchandise store, and sub-forum for the classic E-body Challenger.

Clifford Performance
www.cliffordperformance.net
With the famous slogan 6=8, Jack Clifford started Clifford Research in 1967, extolling the virtues of the inline six. From being a NHRA champ, Jack Clifford's company now offers performance hardware for 4 and 6 cylinder vehicles.

Phoenix Specialty Coatings
phoenixspecialtycoatings.com
Purveyors of exceptional Mopar parts restorations, custom one-offs, and multi-color powder coatings. PSC is owned by Leanna Johnson (a.k.a. The Mistress of Metal) and Billy Lee. Leanna's 1968 Barracuda can be seen at the website.

Endnotes

(Indicated by superscript numbers in the main text)

(1) Barney Sharratt, "Roy Of The Rootes" *Thoroughbred & Classic Cars* (May 1993) p65
(2) "The Crown Comes Back" *CAR Life* (April 1969) p29
(3) Steve Kelly, "Filet of Barracuda – by Barris" *Motor Trend* (April 1967) p90
(4) Julian Balme, "All-American Racers" *Classic & Sports Car* (June 1998) p139
(5) Mark Walton, "Cannonball!" *Thoroughbred & Classic Cars* (December 1999) p120
(6) Ray Brock, "HRM Tests The Dodge Dart" *Hot Rod* (November 1959) p30
(7) Andy Mikonis, "Norm Thatcher. The Grand Old Man Of Speed" *Collectible Automobile* (June 2015) p61
(8) John Bentley, *All The World's Cars, 1954* (New York, Cornell Publishing Corp 1954) p109
(9) Jim Whipple (ed.) "All The 1964 Models" *Popular Mechanics CAR FACTS* (Popular Mechanics Company, 1963, New York) p3
(10) "Four Compact Cars" *Consumer Reports* (February 1980) p114
(11) Julian McNamara, "Mr Bristol" *Thoroughbred & Classic Cars* (April 1994) p67

Also from Veloce ~

Dodge Dynamite – 50 Years of Dodge Muscle Cars
ISBN: 978-1-787110-93-9

This book tells the action-packed story of how the Dodge brand came about, and how its products became the cars to beat. High-octane images show the record-breakers, film stars, race-cars and concepts, doing what Dodge does best: going faster than anyone else, and in superb style.

Also from Veloce ~

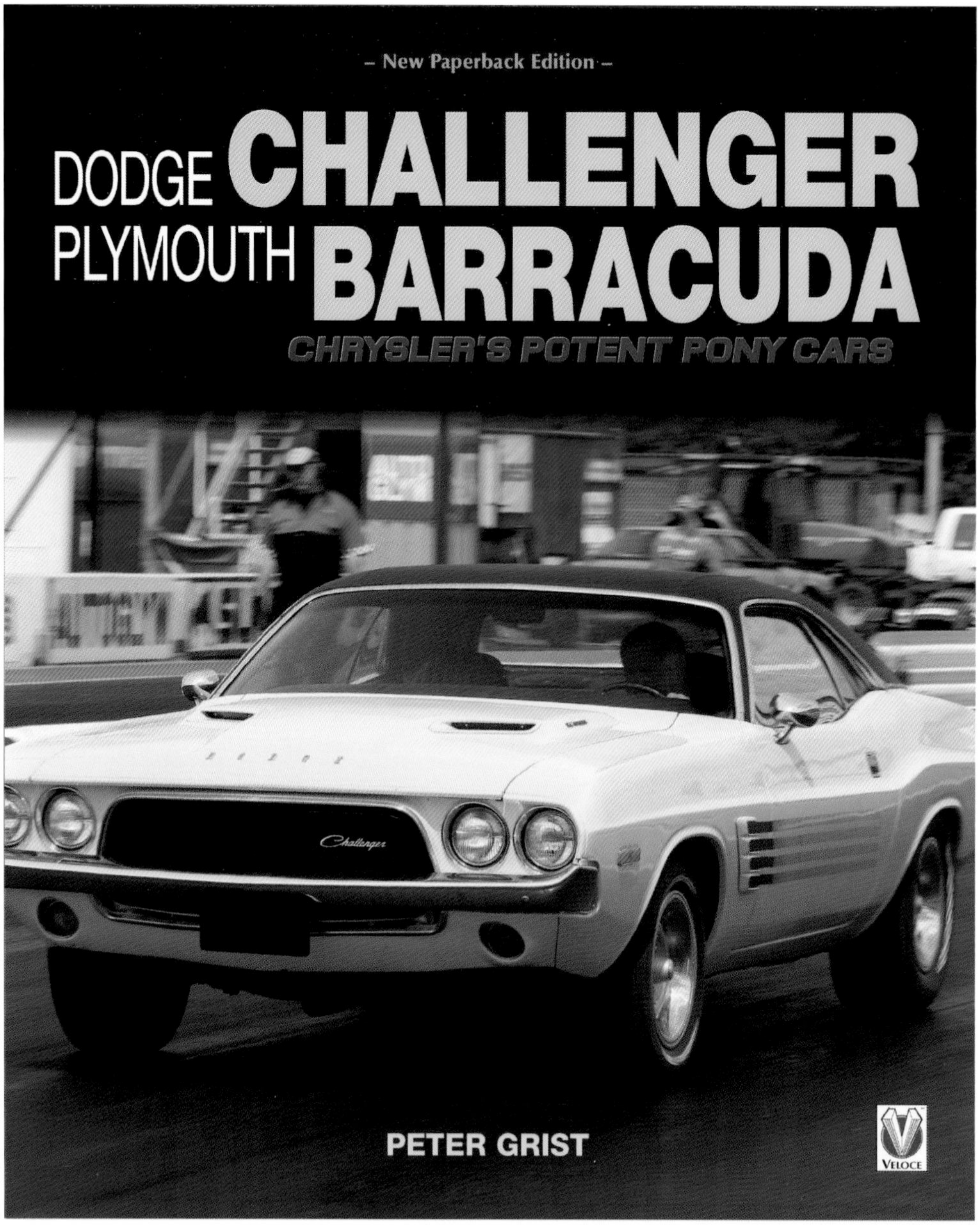

Dodge Challenger & Plymouth Barracuda – Chrysler's Potent Pony Cars
ISBN: 978-1-787110-94-6

Availble again after a long absence! The ultimate Chrysler pony car book written in Peter's now familiar and welcome unpretentious style, featuring every series and model of pony car made by the Chrysler Corporation in the 1960s and 70s, including the slippery Barracuda and classic Dodge Challenger, along with stars of the big screen and drag strip.

For more info on Veloce titles, visit our website at www.velocebooks.com

Also from Marc Cranswick ~

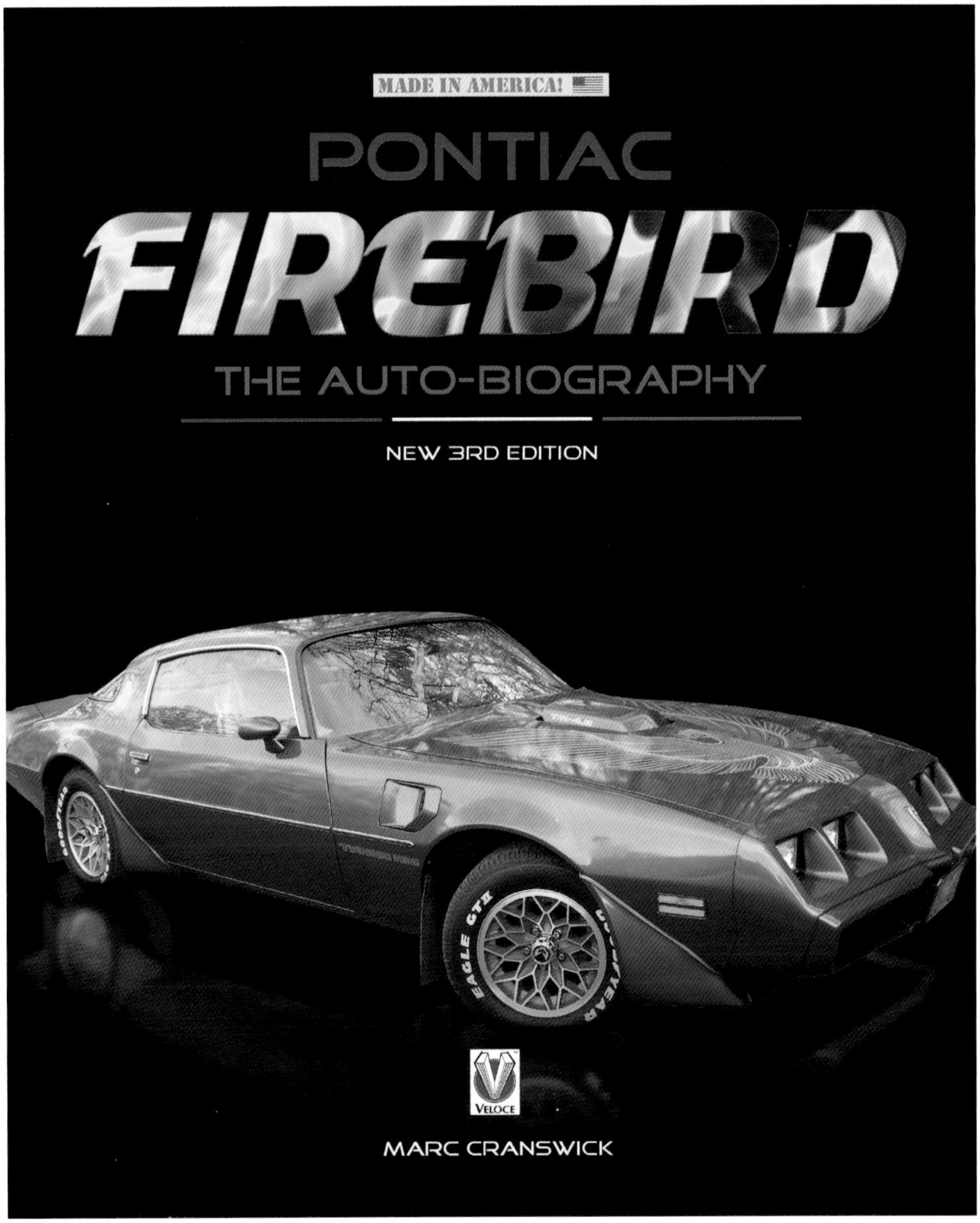

Pontiac Firebird – The Auto-Biography
ISBN: 978-1-78711-003-8

In definitive detail, the story of Pontiac's F-body coupé & convertible during three decades of production & from the height of the personal car/sporty car era of the late '60s, through the fuel crisis/safety-first '70s, '80s renaissance, & '90s indifference. Includes Esprit, SE, Formula & Trans-Am.

• email: info@veloce.co.uk • Tel: +44(0)1305 260068

Also from Marc Cranswick ~

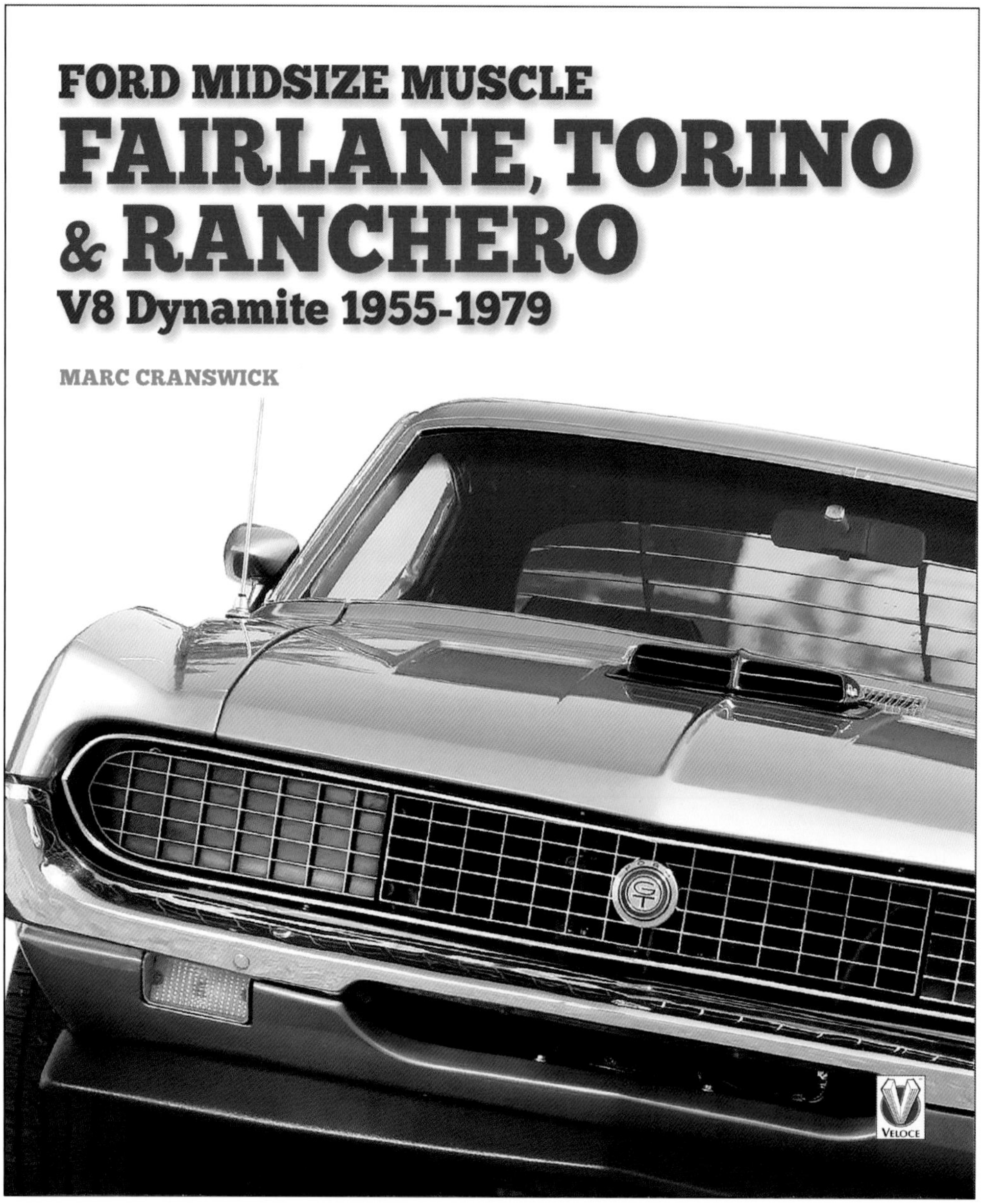

Ford Midsize Muscle – Fairlane, Torino & Ranchero
ISBN: 978-1-845849-29-0

This book details the evolution of Ford's family car through the golden era of Detroit. It tells how Henry took the no-frills Fairlane, added more zing to create the Torino, and satisfied America's luxury desires with the LTD II; and follows the evolution of Ford's midsize muscle cars, to the creation of the first car-based pickup – the Ranchero.

Index

1984 (novel) 78
A38 Police Package 84, 112, 114
A43 Kit Car (F-body) 125, 126, 128
A67 Super Coupe (F-body) 125, 126, 128
AC Ace 155
AC Cobra 155
Adamowicz, Tony 47, 66
Alfa Mito 133
AMAG Automobil und Motoren AG 50
AMC Alliance 132
AMC AMX 21, 29, 40, 79, 81, 125, 126
AMC Concord 117, 126, 130, 131
AMC Gremlin 135, 153
AMC Encore 132
AMC Hornet 117
AMC Javelin 28, 36, 40, 46, 49, 75
AMC Matador Coupe 50
AMC Sportabout 117
American Motors 8, 132
American V8 Road Racing Team 148
Apache 392 V8 138
Arnaz, Lucie 162
Autodynamics 46
Axe, Roy 14

Baker, Steve 94
Baldwin, Maury 88
The Baron (TV show) 159
Barreiros brothers 163
Barris, George 27, 69, 70
BBC Top Gear 135
Beijing Benz 134
Benny, Jack 134
Bertone 163
Bertone Simca Coupe 164
Big Red (Barracuda Formula S prototype) 11
Bixby, Bill 48
Black, Keith 46
The Blues Brothers (1980 movie) 151
BMW 328 156
BMW 507 155
BMW 5 series 125, 135
Bouwkamp, Burt 53, 116
The Brady Bunch (TV show) 48
Branstner, Dick 63, 64
Bricklin, Malcolm 133
Bristol 155-157, 159
Bristol 407 156
Buckel, Mike 57
Bucknum, Ronnie 47
Buick 81, 87
Bullitt (1968 movie) 141

Cadillac Allante 131
Cadillac Seville 93, 131
Caldwell, Ray 46
California Custom (F-Body) 129
Calspan 131
Cameron, Carl 43
The Cannonball Classic 51
Cannonball Run 50, 51
Cannonball Run II (1983 movie) 155
Captive-Aire (runflat tires) 94, 95
Car and Driver 7, 10, 18, 20, 30, 35, 36, 50, 53, 61, 66, 69, 72, 81, 125, 128, 131, 135
Carlsson, Erik 14
Carlton, Don 153-155
Cars and Concepts 128
Castaing, Francois 132
Cerberus Capital Management 132
CGE Pacer Pickup 133, 134
Charipar, Jack 93
Charlie Allen Crenshaw Dodge 64, 71
Chemin, Henri 50
Chesebrough, Harry E 91, 164
Chevrolet Bel Air 52
Chevrolet Biscayne 52
Chevrolet Camaro 14, 16-18, 20, 26, 41, 42, 49, 72, 81, 115, 143, 149
Chevrolet Chevelle 105
Chevrolet Chevy II 72, 100, 101
Chevrolet Citation 121, 130
Chevrolet Corvair 91, 92, 94, 101
Chevrolet Corvette 6, 7, 10, 62, 69, 75, 81, 101, 102, 125, 127, 162
Chevrolet Impala 52
Chevrolet Malibu 106
Chevrolet Monte Carlo 50
Chevrolet Nova 79-81, 131
Chevrolet Vega 121, 153
Chicago Auto Show 144
Chrysler 180 163
Chrysler 300 Mopar '12 (LX) 136
Chrysler 300 Motown Edition (LX) 136, 152
Chrysler 300 Touring (LX) 142
Chrysler 300C 88, 115
Chrysler 300C (LX) 6, 134, 135, 137-139, 142
Chrysler 300M (LH) 132, 136
Chrysler 300S (LX) 136, 137
Chrysler 300SRT8 (LX) 135, 136
Chrysler 300SRT8 Core (LX) 136
Chrysler Airflow 87, 92
Chrysler Australia 130, 155, 163, 164
Chrysler Building (New York) 87
Chrysler C-300 88
Chrysler Charger 164
Chrysler Courier 229 (boat) 163
Chrysler Crossfire 143
Chrysler Espana SA 163, 164
Chrysler Europe 14, 130, 163, 164
Chrysler Fifth Avenue 131, 132
Chrysler Firepower V8 87, 88
Chrysler Imperial 52, 102, 131
Chrysler International Studio 163
Chrysler LX platform 88, 134-138, 140, 142, 146, 152
Chrysler Marine 163
Chrysler-Maserati 131
Chrysler Neon 132
Chrysler New Yorker 88, 131
Chrysler Newport 57, 134
Chrysler Pacifica 143
Chrysler Performance Parts Service 38
Chrysler PT Cruiser 132, 143
Chrysler Royal 163
Chrysler Six 87
Chrysler Valiant 94, 130, 155, 164
Chrysler, Walter P 87
Clean Air Package 63, 67, 104
Clean Air System 104
Clifford, Jack 9, 169
Clifford Performance 9, 99, 169
Clifford Research 169
Clincher Warranty 106, 112
Colbert, Lester Lum 'Tex' 91
Concept 300 (1991) 134
Congratulations, It's A Boy! (1971 TV movie) 48
Contemporary Fiberglass 64
Converse Engineering 155
Cook, Emery 64
Crook, Anthony 156
Crosley 87
Cunningham C4R 51
Curtis, Tony 160

Dallafior, John 94
Daroo I 70
Daroo II 70
Dart (showcar) 52, 89
Datsun Z car 51, 130
Davis, John H 129
Daywalt, Rob 99
De Tomaso, Alejandro 131
Debenedictis, John 148
DeSoto 52, 87, 88
DeSoto, Hernando 87
Detroit Motor Show 143
Direct Connection 128
DLI (Dick Landy Incorporated) 76, 135
Dodge 330 57, 58
Dodge 3700 163
Dodge A-100 66
Dodge Aries 131
Dodge Aspen 6, 86, 116-118, 120, 121, 124-131, 154, 155, 169
Dodge Aspen Custom 118, 124
Dodge Aspen R/T 6, 124-127, 129
Dodge Aspen SE 118, 120, 124
Dodge B-series van 71
Dodge Caravan 131
Dodge Challenger 6, 43-51, 72, 110, 128, 136, 143, 144, 152, 154, 169
Dodge Challenger 570X (LC) 146
Dodge Challenger Deputy 48
Dodge Challenger Legacy by Petty (LC) 147
Dodge Challenger R/T 6, 45-47, 49, 51, 144, 146
Dodge Challenger R/T (LC) 137, 138, 144, 146, 148
Dodge Challenger R/T Classic (LC) 144
Dodge Challenger Rallye 47-51
Dodge Challenger Richard Petty Signature Series (LC) 147
Dodge Challenger SE 45
Dodge Challenger SE (LC) 144, 146
Dodge Challenger SMS 570 (LC) 146
Dodge Challenger SpeedFactory SF600R (LC) 147
Dodge Challenger SRT8 392 143, 146, 148
Dodge Challenger SRT8 Spring Special (LC) 144
Dodge Challenger SRT Demon (LC) 147, 148, 167
Dodge Challenger SRT Hellcat (LC) 138, 141, 144, 146, 147
Dodge Challenger SXT (LC) 144, 146
Dodge Challenger T/A 340 37, 46, 47
Dodge Challenger Western Sport Special 48
Dodge Charger 6, 28, 43, 49, 63, 64, 109, 136-138, 143, 146
Dodge Charger Enforcer (LD) 139, 140
Dodge Charger R/T 138, 140
Dodge Charger SRT8 (LD) 141
Dodge Charger SRT8 (LX) 140, 141
Dodge Charger SRT Hellcat (LD) 138, 141
Dodge Charger Super Bee (LD) 141
Dodge Colt 153, 164
Dodge Coronet 32, 85
Dodge Coronet R/T 150
Dodge Custom 880 57, 58
Dodge D5 Hemi Colt 153, 154
Dodge D-Dart (drag racing) 66
Dodge Dart (compact) 6, 58-68, 70-86, 98, 100, 102, 104, 111-113, 115-117, 128, 149-151, 163
Dodge Dart (fullsize) 52-54, 57, 90
Dodge Dart (midsize) 54-57, 69, 91, 165
Dodge Dart 330 54
Dodge Dart 440 54
Dodge Dart GT 58, 60-64, 67, 69
Dodge Dart GTS 66-73, 149
Dodge Dart SE 81-83, 112, 115
Dodge Dart Charger 63, 64
Dodge Dart Chrome Yellow Taxi Package 86
Dodge Dart Lite 86
Dodge Dart Phoenix 52, 53
Dodge Dart Pioneer 52, 53
Dodge Dart Polara 500 53, 54
Dodge Dart Seneca 52-54, 88
Dodge Dart Sport 77-81, 83-85, 124, 128
Dodge Dart Sport 'Caravan Tan' 85
Dodge Dart Sport 'Hang 10' 85
Dodge Dart Sport 'Spirit of '76' 85
Dodge Dart Swinger 70-74, 78, 106, 111
Dodge Dart Western Sport Special 72, 73
Dodge Daytona Turbo Z CS 138
Dodge Demon 38, 49, 74-78, 81, 108, 147, 148, 167
Dodge Diplomat 123, 129, 131, 159
Dodge Durango 137
Dodge Hemi Charger 500 26
Dodge Hemi Dart (LO23) 22, 24, 56, 64, 68, 69, 70
Dodge Lancer 54, 55, 57-59, 62
Dodge Li'l Red Express 129
Dodge Magnum 128
Dodge Magnum R/T (LX) 137, 142, 143
Dodge Magnum SRT8 (LX) 142, 143
Dodge Matador 52, 54
Dodge Monaco 85
Dodge Omni 129, 130, 132, 138
Dodge Omni GLHS 138
Dodge Polara 52, 57
Dodge Ram Heavy Duty Truck 137
Dodge Shelby Charger 138
Dodge Viper 132
Donohue, Mark 36, 46
Drag Race Package (LC) 144
Duel (1971 movie) 106
Duesenberg 134
Dutra, Doug 'Dr. Dodge' 93, 115
Dyer, Gary 150

E58 Police Motor 125, 126
Eagle Premier 132
Earl, Harley 89
Earls Court Motor Show 162
Eaton, Bob 132
Eckstrand, Elton 57
Electronic Lean Burn (ELB) 123, 126
Electronic Spark Control (ESC) 123
Eminem 135
EVTS (Electronic Vehicle Tracking System) 146
Exner, Virgil 54, 88, 89, 91, 98, 134, 163

Facel Vega 162
Facel Vega II 162
Falcon (Chrysler Corp concept car) 91
Ferrari Daytona 51
Ferrari, Enzo 156
Fiat 500 133, 136, 170
Fiat 600D 102
Fiat 1100-B 102
Fiat Punto 133
Fiat Spyder 102
Fiore, Trevor 156, 157
Fissore 156, 157
Fonzmobile (Duster Coupe) 128
Ford Custom 52
Ford Fairlane 52, 58
Ford Fairmont 121, 129, 130
Ford Falcon 91-93, 100, 101
Ford Galaxy 52
Ford Gran Torino 50
Ford, Henry II 91
Ford Mustang 7, 8, 9, 14, 16-18, 25-27, 29, 36, 38, 40, 42, 43, 50, 62, 64, 83, 129, 143, 154
Ford Pinto 153
Ford Thunderbolt 57, 64
Ford Torino 105, 106
Formula S package 11-18, 20, 30, 62, 102
Formula SX Barracuda 27
Forrest, Steve 160
Franchi, Sergio 116
Frua 156
Fuel Pacer System 81, 86, 112, 113
Fujisawa Shonan Chrysler-Jeep (dealer) 169, 170

Gale, Jeff 143
Garlits, Don 5, 32, 63-65, 138, 146, 153, 169
Garlits' Dart 32, 63-65
General Lee Charger 138, 139, 141
Geneva Motor Show 157, 159, 161
Get Smart (TV show) 163
Ghia 52, 89
Ghia Adventurer 91
Ghia D'Elegance 91
Ghia Gilda 89
Gilles, Ralph 134
Glidden, Bob 153, 154
Golden Commandos 94
Gordy Jr, Berry 135
Grable, Ron 66
Green, Carl 133
Grumman F6F Hellcat 138
Gurney Cougar 18
Gurney, Dan 18, 35-37, 41, 47, 51

Hagenbuch, Pete 121
Happy Days (TV show) 128
Hardy, Don 154
Harmon, Mike 148
Harrison, Rex 116, 117
Harvey, Scott 21, 102
Healey, Donald 162
Heinen, Charles M 63
Hepburn, Audrey 116, 117
Here's Lucy (TV show) 162
Herlitz, John E 15, 28, 43
Hill, Roy 154
Hillman Avenger 163, 164
Hillman Hunter 14
Hillman Imp 163
Hinkle, Jeff 148
Hodges, Clyde 154, 155
Honda Accord 121
Honda Civic 121, 143
Hood, Don 43
Hoover, Tom 57, 152
Hot Wheels 36

Howell, Jay 63, 64
Humber Imperial 156
Huntsville Electronics Division 123
Hurst Firebird 18
Hurst Performance 22, 68, 69
Hutchinson, Pete 34
Hyper-Pak (Slant-6) 59, 71, 91-94, 135

Iacocca, Lee 8, 83, 128, 129, 131-133
IBM 92
iDFSO (Intelligent Deceleration Fuel Shut-Off) 144
Imperial Kustoms 70
Inada, Daijiro 51
International Scout 159, 162
Iso-Clamp (suspension) 118, 122

The Jack Benny Show (TV show) 134
Jeep 132
Jeep Cherokee 132
Jeep Grand Cherokee 137
Jenkins, Bill 'Grumpy' 24
Jennings, Waylon 138
Jensen 157, 159, 160, 162
Jensen 541-S 159
Jensen, Alan 157
Jensen C-V8 159
Jensen-Healey 162
Jensen Interceptor 159, 160, 162
Jensen, Richard 159
Johnson, Don 48
Jones, Parnelli 36

Kaiser-Fraser Henry J 95
Keinener, Esko 102, 103
Keller, Kaufman T 89
King, Alan 116
King, Jake 20, 21, 34, 154
Kobe, Ray 94
Koveleski, Oscar 51
Kraus, Norman 149-151, 154
Kriner, Clint 99
Kriner's Racing Engines 99

Lamborghini, Ferrucio 156
Lancia Thema 136
Landy, Dick 24, 25, 56, 65, 76, 81, 154
Lawrence, Cameron 148
Lorist, Franz 51
Los Angeles Auto Show 142
Lyles, Ronnie 154

Macomb Missile 154, 155
Magna Steyr 134, 136
Makinen, Timo 102, 103
Mannix (TV show) 48, 69, 70
Marchionne, Sergio 133, 136
MARIERIKA 6, 133, 137
Maserati Quattroporte 156
Mason, Jackie 98
Mass, Jochen 50
Martin, Buddy 21, 24, 34
Martin, Dean 116
Matsura, Shunsuke 43
Maxwell, Dick 57
Maxwell Motor Co 134
McEwen, Mongoose 32
McFarland, Jack 70
McLaughlin, RD 22, 86
McNay, Wes 51
MDS (Multi Displacement System) 135-140, 144
Mercer 134
Mercury Capri 129
Mercury Comet Cyclone 65
Mercury Cougar 45
Mercedes 190E 131
Mercedes 600 102
Mercedes E class 31, 135
Mercedes S class 13, 20, 21, 58, 114, 128
Mercedes SL 155
Michigan Police Dept 126
Miller Racing Challenger (LC) 148
Mobile Economy Run 6, 18, 53
The Mod Squad (TV show) 28, 48
Mod Top (Barracuda 1969-70) 28
Monica 162
Monteverdi 123, 156, 157, 159, 162
Monteverdi 375/4 156, 157
Monteverdi 375L 156, 157
Monteverdi 375S 156
Monteverdi 400SS 156
Monteverdi Hai 156-161
Monteverdi High Speed 134
Monteverdi, Peter 156-159
Monteverdi Safari 159, 162
Monteverdi Sahara 159, 162
Monteverdi Sierra 123, 159
Mopar '10 (LC) 146
Mopar Missile (Aspen) 154, 155
Mopar Missile ('Cuda) 152
Mopar Missile (Duster) 152
Mopar Performance 135, 136
Mopar Upfit Center 146
Morris Minor 89
Moss, Pat 14
Moss, Sterling 14
MotorWeek (TV show) 129
Motown Missile (Challenger) 152
Motown Records 136, 152
MPR Racing 146
Mr Gasket 151
Mr Norm's 50th Anniversary GSS 'Cuda Racer Edition (LC) 146
Mr Norm's 50th Anniversary GSS Challenger (LC) 145, 151
Mr Norm's Dart GSS 440 149-151
Mr Norm's Demon 340-6 150
Mr Norm's Demon SS 150
Mr Norm's Dodge Super Charger 150
Mr Norm's Grand Spaulding Dodge 81, 145, 149-151
Mr Norm's Hurst Heritage GSS Charger (LD) 140, 141
Mr Norm's Kawasaki 150
Mr Norm's Sport Club 151, 169
Mr Norm's Super Charged 500 151
Mr Norm's Super Stock Demon 150
My Fair Lady (1964 movie) 116, 117

NASA 152
Nash Bridges (TV Show) 48
Nash, Ron 128
Nationwide Series Competition (NASCAR) 148
New York International Auto Show 135, 136, 148, 164
Newberg, William 92
Newman, Barry 48
NHTSA 131
Nichols, Byron J 58,106
Nicholson, 'Dyno' Don 24, 57
Nielander, Brian 143
Norseman (showcar) 89
Nürburgring 143
Nystrom, Liz 14

Obama, President Barrack 138
Oldham, Joe 78-81
Options (car magazine) 51

Packard 134
Palimedes, Romeo 154
Panch, Marvin 94
Penske (Team) 36, 46
Pentastar V6 135, 141, 144, 146
Perau, Ron 70
The Persuaders! (TV show) 162
Petty Enterprises 128
Petty, Richard 11, 91, 94, 128
Philadelphia Auto Show 144
Pitcock, Forrest 94
Pittsburgh Plate Glass (PPG) 7
Plymouth AAR 'Cuda 340 34-38, 46
Plymouth Advanced Studio 43
Plymouth Arrow 153, 154
Plymouth Barracuda (A-Body) 7-27, 29, 30, 32, 61, 102, 163
Plymouth Barracuda 340-S 14, 18-22
Plymouth Barracuda 383 18, 20, 27
Plymouth Barracuda Charger Power 22
Plymouth Barracuda Coupe (E-body) 28-43, 45, 49, 51, 72, 169
Plymouth Barracuda Gran Coupe 29, 33, 39
Plymouth Barracuda Super Stock (BO29) 20, 23, 32, 34, 56, 64, 70
Plymouth Belvedere 52, 101
Plymouth Cricket 130, 153, 163, 164
Plymouth 'Cuda 340 (1969) 14, 20, 25, 26, 43, 125
Plymouth 'Cuda 383 (1969) 25
Plymouth 'Cuda 440 (1969) 25, 26, 31, 71, 166
Plymouth 'Cuda 340 (E-body) 6, 30, 32-34, 38-42, 74, 76, 126
Plymouth 'Cuda 360 (E-body) 43
Plymouth 'Cuda 383 (E-body) 6, 30-34
Plymouth 'Cuda 440-4 (E-body) 6, 31, 33, 38
Plymouth 'Cuda 440-6 (E-body) 6, 28, 30, 31-34, 38, 39, 41-43, 166
Plymouth 'Cuda Hemi 426 (E-body) 5, 26, 31-34, 36, 38, 39, 41, 48, 50, 152, 166
Plymouth Duster 6, 38, 49, 73, 74, 76, 78, 105-112, 115, 118, 122, 124, 128, 137, 152, 153, 167, 168
Plymouth Duster Twister 106
Plymouth Feather Duster 86
Plymouth Fury 18, 19, 52, 56, 101, 115
Plymouth Gran Fury 131
Plymouth Horizon 129-132
Plymouth National Barracuda Club 14
Plymouth PA 130
Plymouth Prowler 132
Plymouth Reliant 131
Plymouth Road Runner 21, 32, 72, 105, 129
Plymouth Satellite 32, 106
Plymouth Savoy 52, 57, 88, 89, 90, 101
Plymouth Superbird 43, 51
Plymouth Valiant 6-11, 13, 16, 54, 55, 57, 61, 87, 91-116, 131
Plymouth Valiant Brougham 112
Plymouth Valiant Red Snapper 102
Plymouth Valiant Scamp 111, 112
Plymouth Valiant V100 95-97
Plymouth Valiant V200 95-98, 100, 103
Plymouth Valiant V200 Signet 96-98, 100, 101, 103, 104
Plymouth Volaré 6, 116-122,124-129, 131, 169
Plymouth Volaré Custom 118, 119, 122, 124, 129
Plymouth Volaré Duster 118, 122, 125, 129, 131
Plymouth Volaré Front Runner 125
Plymouth Volaré Fun Runner 125
Plymouth Volaré Premier 118-120, 122, 124, 125
Plymouth Volaré Road Runner 6, 124, 125, 127, 129
Plymouth Volaré Suburban (wagon) 120, 121
Plymouth Volaré Sun Runner 125
Plymouth Volaré Sunrise 125
Plymouth Voyager 131
Polka-Dart 63, 64
Pomeroy, Stewart 152
Pontiac Firebird 29, 34, 40, 41, 43, 45, 49, 79, 81, 125, 127
Pontiac Grand Am 40
Pontiac GTO 76, 81
Porsche 911 58
Posey, Sam 37, 46, 47
PowerFlite (automatic) 87, 91
PRDA (Polish Racing Drivers of America) 51
Project A901 91
Project 929XC 163
PSA Group (Peugeot-Citroen) 130, 132, 164

Quiet Car Package 77, 81
Qvale, Kjell 162

Radio Suppresion Pack 86
Rallye Package (LC) 144, 146
Rallye Redline (LC) 146
Ram Induction (D-500/Super Red-Ram V8) 53-57
Rambler American 8, 58, 89, 94, 101
Rambler Tarpon 8
Ramcharger 413 pack 55-57, 69, 165
Ramchargers 57, 94, 150, 152
Range Rover 159
Renault Dauphine 90
Renwal 134
Revival Cars 134
Ricketts, Mike 155
Rootes, Billy (Lord Rootes) 163
Rootes Group 14, 163, 164
Rootes, Sir Reginald 156
Rover 132
Roy Woods Racing 128

Sampson, John 7, 15, 43
Sarafian, Deran 48
Sarafian, Richard C 48
Savage, Swede 35-37, 46
Scat Pack 70-73
Schirmer, RM 'Ham' 116
Sheen, Charlie 48
Shelby, Carroll 138, 155
Shelby Mustang 18, 138
Simca 50, 130, 163, 164
Simca GLS 163
Simenson, Troy 94
Sinatra, Frank 131
Smith, Carroll 46
Snoop Dogg 135
Society of Illustrators 88, 98
Sox, Ronnie 20, 21, 23, 24, 34, 41, 100, 153, 154
Sox & Martin 20, 21, 23, 24, 34, 154
Spehar, Ted 153
Spelling, Aaron 48
Spielberg, Steven 106
Sport Wagon (F-body) 129
Sports Accessory Group 20
Starr, Ringo 162
Stevens, Joe 148
The Stoolie (1972 movie) 98
Studebaker Lark 89, 94, 101
Stutz 134
Sullivan, Ed 116
Sunbeam Alpine 164
Sunbeam Fastback Coupe 164
Sunbeam Imp 164
Sunbeam Rapier 14, 163, 164
Sunbeam Tiger 163, 164
Super Six 225 118, 121, 123
Super Stocks (drag racing) 20-25
Sure-Grip (lsd) 12, 13, 17, 22, 33, 47, 60, 70, 101, 102, 104, 111, 119, 121, 165, 166
Swig, Martin 51
Swingle, Connie 64
SXB Barracuda 27

TA2 Series (Trans Am) 148
Takao, Erika 5, 136, 137, 170
Takao, Marie 5, 136, 137, 170
Talbot Tagora 130
TARP (Troubled Asset Relief Program) 132
Tastevin, Jean 162
Taylor Motor Sales (dealer) 57
Terminal Velocity (1994 movie) 48
Thatcher, Norm 50, 57, 105
Thompson, Dean 154
Thornton, Jim 57
TJ Hooker (TV show) 131, 143
Touring 160
Truffaut, Francois 162
Tullius, Bob 66

Vanishing Point (1971 movie) 48, 51
VM Motori 136
Volvo 140/160 29
Volvo 240 155
Volvo 940 155
Volvo PV series 130
VVT (Variable Valve Timing) 138-140
VW Beetle 59, 78, 90, 92, 98, 102, 106, 110, 130
VW Karmann Ghia 92, 163
VW Rabbit 25, 121, 131

Weaver, Dennis 106
Weiner, Larry 141, 151
Werley, John 57
White, Derek 148
Willys-Overland 87
Winkleman, Otto 30
WLS Chicago 149
Wylie, Frank 63-65

Xfinity Series (NASCAR) 148

Yates, Brock 50, 51
Yates, Brock Jr 50, 51
Yeleg, JJ 148
Yenko Camaro 18
Yenko Stinger 92
Yetter, Gene 152
Yowell, Bobby 154

Zagato 156
Zetsche, Dieter 135